Innovation Generation

Also available from ASQ Quality Press:

Do It Right the Second Time, Second Edition
Peter Merrill

How Organizations Learn: Investigate, Identify, Institutionalize
Patrick L. Townsend and Joan E. Gebhardt

The Executive Guide to Understanding and Implementing Lean Six Sigma: The Financial Impact
Robert M. Meisel, Steven J. Babb, Steven F. Marsh, and James P. Schlichting

Transactional Six Sigma for Green Belts: Maximizing Service and Manufacturing Processes
Samuel E. Windsor

Strategic Navigation: A Systems Approach to Business Strategy
H. William Dettmer

Managing the Customer Experience: A Measurement-Based Approach
Morris Wilburn

Lean Kaizen: A Simplified Approach to Process Improvements
George Alukal and Anthony Manos

Root Cause Analysis: Simplified Tools and Techniques, Second Edition
Bjørn Andersen and Tom Fagerhaug

The Certified Manager of Quality/Organizational Excellence Handbook, Third Edition
Russell T. Westcott, editor

Enabling Excellence: The Seven Elements Essential to Achieving Competitive Advantage
Timothy A. Pine

The Certified Six Sigma Green Belt Handbook
Roderick A. Munro, Matthew J. Maio, Mohamed B. Nawaz, Govindarajan Ramu, and Daniel J. Zrymiak

Leading Peak Performance: Lessons from the Wild Dogs of Africa—How to Create Pack Leadership and Produce Transformative Results
Stephen Hacker and Marvin Washington

To request a complimentary catalog of ASQ Quality Press publications, call 800-248-1946, or visit our Web site at http://www.asq.org/quality-press.

Innovation Generation

Creating an Innovation Process and an Innovative Culture

Peter Merrill

ASQ Quality Press
Milwaukee, Wisconsin

American Society for Quality, Quality Press, Milwaukee 53203

Printed in the United States of America
14 13 12 11 10 09 08   5 4 3 2 1

Library of Congress Cataloging-in-Publication Data

Merrill, Peter.
Innovation generation: creating an innovation process and an innovative culture / Peter Merrill.
p. cm.
Includes bibliographical references and index.
ISBN 978-0-87389-734-1 (hbk. : alk. paper)
1. New products. 2. Creative ability in business. 3. Technological innovations. I. Title.

HF5415.153.M47 2008
658.4'063—dc22   2008011887

ISBN: 978-0-87389-734-1

Publisher: William A. Tony
Acquisitions Editor: Matt T. Meinholz
Project Editor: Paul O'Mara
Production Administrator: Randall Benson

ASQ Mission: The American Society for Quality advances individual, organizational, and community excellence worldwide through learning, quality improvement, and knowledge exchange.

Printed in the United States of America

 Printed on acid-free paper

Quality Press
600 N. Plankinton Avenue
Milwaukee, Wisconsin 53203
Call toll free 800-248-1946
Fax 414-272-1734
www.asq.org
http://www.asq.org/quality-press
http://standardsgroup.asq.org
E-mail: authors@asq.org

To my mother Phyllis Merrill, a remarkable lady, who taught me tenacity, the primary skill of the innovator.

To my daughters Rachel and Sarah, today's "innovation generation," who keep my ideas fresh.

To Angela, the other half of my life, whose creativity balances my focus.

To my late father William and grandfather Herbert, both of whom were writers, artists, and scientists.

Table of Contents

Part II The Process

Part III The People

Preface

My grandfather and father were both artists. My grandfather excelled in pastels and my father in watercolors. They were also both writers. As each weekend approached, my father would sit down and write his sermon for the coming Sunday. Whenever we visited my grandparents I would see my grandfather doing the same. On Sundays, both morning and evening, my father, a lay preacher in the Methodist church, would deliver his sermon. My grandfather was a chemist before he entered the church and my father was an electrical engineer, maintaining his membership in the Institute of Electrical Engineers until he died.

With those influences it is little surprise that I am a writer, an artist, and an engineer, and that I give a lot of my time to keynote speaking.

However, there is a creative component inside me that comes both from having started painting as a child and having been given a chemistry set at a young age. Chemistry took me down the road to become a chemical engineer, but in truth what I enjoyed about chemistry was the incredibly creative and visual and sensory aspects of the chemistry lab. I guess that is where the desire to experiment began.

My exploring started in my teens. At the age of twelve I bought my cousin Bill's bicycle and started going places I had never gone before. Travel was exciting. At the age of fourteen I cycled from the Midlands of England to the Lake District with three of my friends, Dave, Bob, and Roger, and we were away for two weeks. I can't believe my parents allowed it to happen. We stayed in youth hostels and climbed hills and met people. The travel bug bit and has stayed with me my entire life.

When I graduated from Birmingham University, I knew chemical engineering was not the right fit. I was good at it, but I did not have the passion for it. The only part of my degree that still sticks is my thesis. It was on the use of "The Plasma Jet As a Chemical Reactor." Plasma had been discovered as the fourth state of matter and there was widespread excitement at its possibilities.

I still have my original thesis after all these years and interestingly my primary finding as a raw undergraduate was that the science of plasma had been held back through a lack of collaboration between the players. Huge lesson for today's innovator.

From university I joined the Courtauld's Group because my instinct told me they might provide an escape route for me. As well as chemical engineering, Courtauld's was a textile company. However, they first reignited that travel bug as they sent me to places like the Allegheny Mountains in the U.S.A., the Usutu River in Swaziland, and the frozen north of Finland and Sweden. This totally unsettled me, and when I had the chance to "jump ship" from chemicals to textiles, I did so and made the dramatic house move from Coventry to Manchester, England, 150 miles away.

Textiles is an exciting world. I actually left Courtauld's for three years and ran my own fashion business, even though my background was in engineering. In a way, I never left the company because my friends were working there. I still see my friends from Courtauld's. We have the common values that are fundamental to any company with a strong culture. I met other friends in my new home in Manchester through the game of rugby. Both "networks" brought that variety and stimulation that are vital to life.

I rejoined Courtauld's after three years in fashion, and within eight years became chief executive of Christy. I then entered one of the most exciting periods of my life.

When I took responsibility for the Christy brand I inherited 150 years of history. William Miller Christy invented the humble towel as we know it today. Although it is 150 years old, the Christy brand for which I became responsible was an example of breakthrough innovation. Its invention followed the classic stages of innovation. Although I will talk about this in a later chapter these stages are:

The exploration and the opportunity (The grand tour and Victorian culture)

The connection to the solution (The sultan's harem and the Turkish bath)

Engineering and development (The Droylsden factory)

Execution (The Great Exhibition of 1851 and the Royal Turkish brand)

I learned these vital lessons: 1) Even though the Christy brand was 150 years old, you can't stop innovating; 2) innovation does not have to be high in technology; 3) artistic and scientific innovation follow the same process.

Christy was a great learning experience for me and I will refer to it several times in this book. However, after three years I was presented with an opportunity that few would decline.

I left Courtauld's and everyone was stunned. I was regarded as a company man. I had the opportunity to work with Phil Crosby, and entered another period of excitement and change. After I had been with Phil for close to five years, he retired, and I decided again to move on.

I set up my own consulting practice, which I have now had for over fifteen years. During that time I have had the privilege of working with some of the most exciting people and companies on the planet. I have helped organizations develop their management systems by using the ISO 9001 framework, and I have at the same time become very involved in the ISO organization. I have represented Canada on the ISO/TC 176 and I have also been on the strategic advisory group of ISO/TC 176. This role is honorary, and people often ask me why I pay out of my own pocket to do this; I see ISO as one of those global forces that are making the world a better place.

I have been to developing countries like Kenya, Panama, and Egypt, and seen the remarkable effect that ISO 9000 and ISO 14000 have had in helping those countries develop their economic infrastructure.

Following the instigation of my daughter Sarah, who is HR manager with a major British corporation, I have been able to personally initiate the development of the "People Involvement Standard." These are very fulfilling and rewarding achievements.

Throughout the time I have had my consulting practice I have also been closely involved in the American Society for Quality (ASQ). ASQ is unique in the world as a professional body. It is not elitist and embraces all. I have twice had the privilege of chairing the Toronto section and have spoken at ASQ conferences both nationally and locally. The networking and sharing of knowledge that is facilitated by ASQ is fundamental to innovation.

The other exciting aspect of running a consulting practice is the wide variety of organizations I have been able to work with. Financial institutions like AIG, design organizations like IBM and RIM, manufacturing companies such as Solectron and Husky, and other companies in the food, retail, and distribution industries.

As I've helped these people develop their management systems, the natural question once the system became effective was "Where to next?" Far too many companies just play with their internal system instead of looking outside. Over the last five years I have been drawn into the need for these companies to be more innovative, and my background in both artistic and engineering design has made a perfect fit. I have keynoted on innovation at conferences and trained and consulted with organizations seeking to develop their innovative ability.

One of the key aspects of the work I have done with all of these organizations is the important balance of *people* and *process*. This was reflected in my first book *Do It Right the Second Time* and you will see it reflected in this book as well.

When I talked with Matt Meinholz at ASQ Quality Press about the title of this book and proposed *Innovation Generation* he asked, "Is that a noun or a verb?" My answer was "It's both, actually." The "actually" word is my British heritage coming through and the "both" word takes you back to my first book *Do It Right the Second Time,* which focuses on the importance of balancing both people and process. One other carryover from *Do It Right the Second Time* is the "Browser's Briefing." Readers of that book have all told me how useful it was. At the end of each chapter you will see the Browser's Briefing, which gives you the key points from that chapter.

We truly are an "innovation generation" (the noun) and "generating innovations" (the verb) is one of life's most rewarding experiences.

The world of innovation is exciting. Welcome to the future.

Peter Merrill
Kilbide, Ontario, Canada
March 9, 2008

Acknowledgments

My sincere thanks are due to some of the people with whom I have networked and who have allowed me to mention them in this book:

Angela McCauley, Phil Crosby, David Chater, Bob Lessels, Roger Harle, Sarah Blake, Gary Cort, Armando Espinosa, Phyllis Merrill, Herve Mignot, Deborah O'Leary, Rachel Thomas, Bob Hairsine, Philip Thomas, Liz Thomas, Bill Jappy, Teri Yanovitch, Jim Laforet, Trevor Smith, Lucy Stange, Lorraine Robert, Andy Alltree, Brendan Webb, Anne Wilcock, Paul Simpson, Renato Lee, Lillemore Harnell, Ken Bales, Bill Truscott, Niall Ferguson, Nora Camps, Ian Oliver, Chris Hakes, and Gerry Kavanaugh.

A special thanks is due to Lyn Underwood who typed the tens of thousands of handwritten notes that were the raw material of this book.

Introduction

A BRIEF HISTORY OF INNOVATION

If you look at the 'spikes' in knowledge growth through the recorded history of the Mediterranean, Europe, and North America, these spikes are getting closer together. Ancient Egypt 'spiked' in 3000 BC and the effects of that knowledge explosion lasted for more than half of our recorded history, until 1000 BC. The Greco-Roman knowledge explosion spiked with Alexander and Aristotle in 400 BC. This lasted less than 1000 years, till the fall of Rome in the fourth century AD.

We emerged from the dark ages with the Renaissance and the work of giants such as Leonardo da Vinci and Michaelangelo. A mere 300 years later came the industrial revolution, and the next big spike in knowledge growth was between 1880 and 1920 with the era of the "inventor." As the United States emerged from the death and carnage of the Civil War and people refocused on living, communities across the country started to grow, and the industrial revolution gathered pace along with business and trade.

People needed to travel to other communities to trade, or better still, talk to people in other communities without the traveling. They needed to spend time in the evening with their families as their work increasingly took them away from home.

Henry Ford saw that the need for speedy travel would not be solved with faster horses, Alexander Graham Bell saw that the need to communicate was not being solved with the telegraph, and Thomas Edison saw the shortcomings of oil and gas lighting in the home.

Just 100 years later we are in the "innovation generation" and another explosion of knowledge through information technology.

We are in an age where people are expecting us to innovate. It's not just nice to come up with new ideas—customers and consumers are expecting it, and are actively "chasing cool."

Unfortunately, business has not positioned itself well for innovation. In the last years of the 20th century, business focused on improving the delivery of existing products and services and less on the development of new markets and new products. As a result, quality management has frequently become internally focused. The harsh reality is that if a product or service has become outdated, effort to improve its delivery efficiency is totally wasted.

Innovation is about developing the products and services that the market needs tomorrow, and is driven by the need for convenience, not by technology. Finding a "cool new idea," while it may be interesting and exciting, has no value unless the idea solves a real problem. Even then, new ideas will only be adopted if they are easy to adopt.

Contrary to popular belief, new ideas emerge most frequently as the result of collective knowledge and typically do not come from a lone "genius." Consequently, for successful innovation to occur the internal and external networks of an organization have to be well developed. An innovative organization enables information to flow freely between its people and also enables information to flow freely to and from external organizations. This type of "networked" organization transmits information and knowledge rapidly and effectively.

Interestingly, documented knowledge typically accounts for only 20 percent of the knowledge in an organization. Good *knowledge management* (KM) is the platform from which innovation is developed, and an innovative organization enables the rapid transfer of the knowledge that is in people's minds by using its network. To innovate we have to release the knowledge in people's minds, and "communities of innovation" are an essential technique for doing this.

We all have a role in the innovation process. *Creators* generate opportunities, *connectors* link opportunities to solutions, *developers* make solutions practical, and *doers* implement solutions.

Through the self-assessment in this book, you and your colleagues can identify where you make your best contributions to innovation. You will find that creators and doers are practical, while connectors and developers are thinkers. The process works best with a different mix of each of these attributes at each stage in the innovation process.

Start your innovation process with core customers, then move on to "not yet" customers. Engaging people in a good innovation strategy will find new markets where there is no competition. However, you must position yourself quickly so that the competition later sees no value in entering your market. "Open market" innovation is then a way of moving us toward people who are *not yet* customers, and this will also help us access that vital knowledge "outside the box." Open market innovation recognizes that there is more knowledge outside the box than inside.

THE PROCESS

Many people are surprised that innovation follows a defined process and that creativity is only part of the innovation process.

The process is driven by need and not by technology. Surprisingly, the market may not realize that the need exists. You need to ask customers questions like "What are your hassles?" and "Where do you waste time?" and also remember that it is easier to change your product than to change your customer base. This "creation" work takes time, and breakthroughs usually occur after a lot of hard work, even though the mode of operation is loose at this early stage. Brainstorming with customers is one of the prime techniques for finding these opportunities.

When we move to the solution stage of the process we find that "connecting" to a different environment frequently reveals a "radical" solution, and we have learned, as Linus Pauling said, "The best way to get a good idea is to get lots of ideas." Using collective knowledge is the most powerful way of finding alternative solutions.

Most of the solutions you find are built on a lot of previous experience and an "epiphany" is really the last piece of the jigsaw puzzle. However, there is one other "pain point" you must go through. Somebody will have to say "You can't do that" when you offer a radical solution.

The tipping point in the innovation process is where we select which options to pursue but, sadly, choices are often made based on poor data, and organizations frequently try to pursue too many options. Organizations are also drawn too often into safe and minor innovations; they have become "risk averse," and only select short-term choices.

A new-product portfolio must include a number of long-term and potentially major innovations in order for an organization to have a healthy future. Risk-taking is fundamental to innovation.

The selection stage or tipping point is also the point where the process switches from "loose" to the "tight" mode required for the development stage, and organizations often have difficulty with this essential behavior change.

The primary challenge at the development stage is to make the new product or service easy to use. To most people's surprise, you will generally need an exponential improvement in time or cost savings for rapid acceptance of a new product. Partnerships with the supply chain and delivery chain are also vital here. At this development stage "time is of the essence." The chance is that someone else on the planet has already found a solution that will compete with your own!

Executing the final working solution is the toughest challenge of all, and research by the Conference Board has shown that of 3000 original ideas only one survives. Engaging operations and sales people in the development stage prepares them for this execution stage. The best ideas don't always make it, and competitors with inferior offerings will copy you if you are not fast to market. The "value proposition" has to link to the customer's previous experience and show just two or three key advantages for your new product. This way, the customer will "get it."

THE PEOPLE

You can see that *collective knowledge* is fundamental for innovation, and innovation succeeds through *communities of innovation* where success will come from the coordinator of the community having "passion for people."

Although the community will be in "stay loose" mode at the outset, it is essential for maintenance of its morale to achieve an early, visible result. The community changes to a "tight" mode of behavior after the tipping point, and team membership may also shift after the tipping point. For a small business, where membership can not change, behavior will need to change.

An innovative culture has to be strong to be effective but must be based on the values of innovation. People's behavior is based on their values, and innovative behaviors are those such as exploration, observation, collaboration, and experimentation. Embracing failure is also a critical behavior.

Networking is an essential behavior for the innovator. Famous people in history often attribute a large measure of their success to their network, as well as their personal ability. Networking is moving from being an art to being a science. Breakthroughs occur at the intersection of different "bodies of knowledge" and so networks need to be diverse and dispersed for knowledge creation.

A business leader plays a key role in communicating these values through a vision of innovation.

IMPLEMENTATION

The organizational assessment in this book will enable you to identify the areas of weakness in your organization and develop an innovation action plan. As the innovation process starts working effectively, the organization should then move to open market innovation. Knowledge lifetime shrinks in periods of rapid innovation, and radical innovation is achieved more easily with ideas from outside your organization.

A leader's most immediate task is to set a strategic direction for innovation, and a leader has to make tough decisions on which ideas to pursue. Strategic direction is derived from market need and the core competencies of the organization.

To make all of these changes happen, a group of "change agents" needs to create a road map and a project plan for change, and you start the change by focusing on a product, service, or market segment with declining revenue.

That early win is vital in order to light the fire of innovation and join the innovation generation.

Part I

The Essentials

1

The Why, What, and How

The future is very important; it is where we will spend the rest of our lives.

—Joel Arthur Barker

It was midsummer and I was visiting my family in England. I had just left the Midlands and I was driving on the M40 motorway between Oxford and London. It was a beautiful clear day, and then in the corner of my eye I saw a small red light start flashing. I glanced down at my BlackBerry device, which was in its familiar resting place. The BlackBerry was a gift from Gary Cort, VP Quality at Research In Motion (RIM), and my good friends at RIM. I had a message.

I quickly read the message on the screen. It was quite simple. My friend Armando Espinosa had e-mailed me from Mexico City and was asking if I could speak at an upcoming conference in Mexico. He wanted me to speak on "Quality Management and Innovation." I returned the BlackBerry to its location and, doing some mental arithmetic, I worked out that the conference would be on Labor Day weekend. I decided that I would pull off at the next exit, which would be Milton Keynes, and confirm with Armando that I would be delighted to accept his invitation.

As I drove the ten miles to the exit I thought to myself how this remarkable little event was a testament to modern ingenuity and innovation.

A guy in Mexico City can send a message to a guy who he thinks is in Canada, but is actually in England, and 10 minutes later will get a positive response. That simple little device, the BlackBerry, had enabled all of this to happen. It is this "world shrinkage" that is such a key factor in the innovative world of today.

Innovation is exciting. It is a new world. Innovation is about developing the products and services that the market needs tomorrow. It is about creating the future. As Joel Arthur Barker said, "The future is very important; it is where we will spend the rest of our lives."

For business success in that future world we must be thinking about tomorrow's customer and what their needs will be. Much of business has tended to focus on today's customer and address their immediate concerns. Business has also focused inwardly, and this has caused us to neglect the future. In the last half of the 20th century, business focused heavily on improving delivery of its products and services. This focus was supported by investment in just two of the four main business process areas: infrastructure, through investment in information technology, and product delivery, through investment in quality management.

Referring to Figure 1.1, the four primary process areas of a business are 1) market development, 2) product development, and 3) product delivery, which are held together by 4) infrastructure.

You can see that for a balanced business that is expecting to have long-term success, the investment in delivery and infrastructure needs to be balanced with investment in market development and product development. That means investment in innovation.

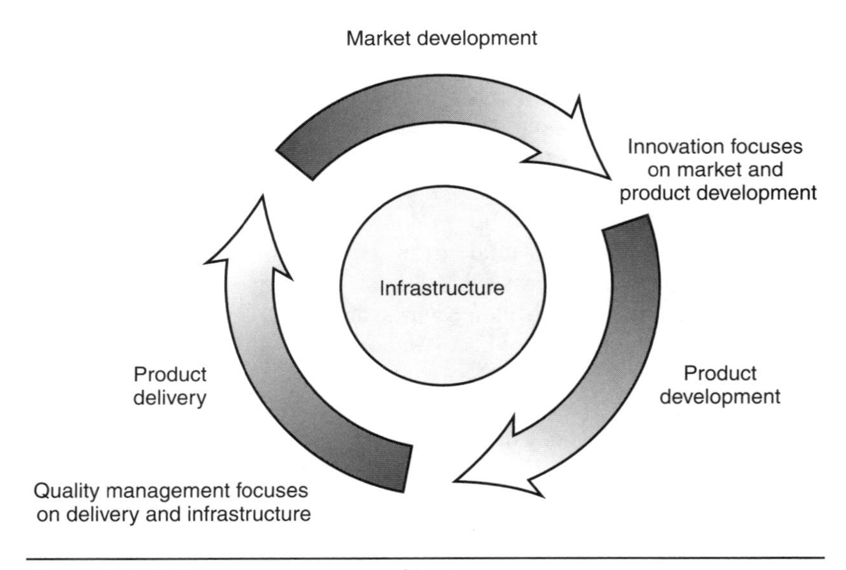

Figure 1.1 Primary process areas of business.

Quality management talks about customer focus, but much of it has become too internally focused and needs to move back to a market focus. In the world of quality management, people have been schooled over the last thirty years to "do it right the first time." However, if you are not *doing the right things* and your product or service has become out of date or "mature," then your effort to improve efficiency is wasted.

Innovation is about developing the *right things*, the products and services that the market needs tomorrow.

A good innovation strategy will put you ahead of the competition, in a future position where the competition sees no value in entering your market. The mighty Microsoft was unable to unseat Intuit and its Quicken software, Cirque du Soleil has created a unique market niche by innovating the tired concept of the circus, and the BlackBerry has established an enviable market position, in spite of attempts by its rivals to displace it. The reason is its remarkable software. Companies chasing Apple arrive at the market space and find that Apple has moved on. These are outstanding innovators. This book will explain how these and other similar companies executed their strategy

Innovation is a critical component in any overall business strategy, and it was Michael Porter who brought strategy into fashion.[1] His "five forces" model is simple. His critical insight is "corporate strategy should meet the opportunities and threats that exist in the external environment." Porter identified five competitive forces that shape every industry and every market. They are the threats from *suppliers, customers, new entrants, substitutes,* and *rivals.*

Innovation is about addressing these threats.

If we look at Porter's work on strategy, his competitive forces *suppliers* and *customers* actually become the allies of the innovator. I will explain this in Chapter 6, "Creating the Opportunity" and then later in Chapter 18, "Open Market Innovation." In periods of rapid knowledge growth such as we are experiencing today, there is more knowledge "outside the box." You will be able to access this knowledge most easily by sharing your own knowledge with your business partners.

Staying with Porter, if you think competitive threats are only on the home front consider the "hidden dragon." Fifty percent of U.S. growth in 1999 was from products less than 10 years old. Many of these products were developed in Asia. If you think low cost is the threat, then think again. It is low income, not low cost, in China and India that drives innovation. India developed the world's largest mobile phone market at 20 percent of Western cost, and China is responsible for more than half of global production of motorcycles. These results were achieved through innovative approaches to service delivery and innovative process networks. North America needs to ask itself, "What is next after the Japanese automobile threat?"

CREATING THE FUTURE

Peter Drucker said the best way to predict the future is to create it. This is true within the context of the mega trends that are always around us. Recognizing trends creates the best opportunities for innovation whether these trends are globalization, information technology, global warming, terrorism, or an aging population. New trends create new needs. New needs create new problems, and so create new opportunities for the innovator.

At the same time this doesn't mean that every innovation has to relate to these mega trends. Within your own world there are many opportunities for innovation.

Innovation is about developing the right things, the products and services that the marketplace truly will need tomorrow.

THE MIND OF THE INNOVATOR

The popular perception of the mind of the innovator is that of the person who finds a cool new idea. Unfortunately there have been many so-called innovative answers or solutions to problems that turned out to be of little significance in the marketplace.

People ponder on a problem that impacts them personally and go through a "connecting" experience. They have that "Aha!" moment or epiphany. The excitement of the epiphany leads them to believe that the problem was more important than they had thought. They have been thinking about the problem for a long time so it has increased its importance in their mind. In truth, they find the answer to a problem, and yet the more important question is "How important is the problem?"

At the other end of the process, having found a potential solution, people all too often forget about the user of the solution. They rush to market or initiate a new service or product and upset the customer with an offering that is full of glitches and bugs. It's like the old days of buying a new car, or today buying new software.

Making an innovation user-friendly is a vital step in the process. If you don't, you will upset the market, and you will rarely get a second chance. More likely, a competitor will see your failure, address your shortcomings, and take your opportunity away from you.

CONVENTION AND INNOVATION

Business wisdom says listen to the customer, that technology drives change, and that you should recognize success. Convention also says target larger markets, seek higher margins, and avoid failure.

These principles all change for the successful innovator. Your market doesn't yet exist, so you must anticipate the market ahead of the customer. You must understand that market convenience, not technology, drives change. Additionally, you must plan for failure and embrace it.

We are surrounded by the results of convenience driving innovation. Photographic film has all but given way to digital photography. Notebook computers are used far less while traveling with the spread of handheld devices such as the BlackBerry. Greeting card shops are under threat as free Internet cards become available. Maybe in 10 years electric companies will be downsizing as solar cells and fuel cells replace them.

I recently had arthroscopic surgery on my knee, the consequence of playing football for too many years. I was able to go into hospital for day surgery and walk to my car afterwards. In the past, this surgery would have required a week in hospital, a month on crutches, and several months of low mobility. I like this kind of convenience.

There are many myths surrounding the world of innovation.

It is often thought that innovation comes from a sudden stunning insight, but if you look at history, most innovations build on a lot of previous experience. People also believe that ideas come from a lone genius whereas in fact most innovations are the result of collective knowledge.

People think that good ideas are hard to find and that they arrive randomly. In fact there is no shortage of ideas. The bigger challenge is identifying the associated opportunities.

The final reality of innovation is that commercialization is the biggest challenge, and most new ideas struggle to see daylight because potential users resist anything new. In truth, the best ideas don't always make it. You will need to have developed an effective quality management system if you want to deliver your new product on time and to customer requirements.

Some people also believe that there is no innovation process, whereas the reality is that many innovations actually fight their way into existence through a process that is so poorly defined people don't realize it exists.

THE PROCESS

Phil Crosby used to say "All work is a process," and yes, innovation is a process.

Innovation starts with identifying a customer or market need. Importantly for the innovator, neither the customer nor the market may recognize that need. Henry Ford said, "If I had asked my customers what they wanted they would have said 'faster horses'." Creative thinking is usually required at this first stage. I have had a fortune cookie note on my notice board for years that says "The secret of a good opportunity is recognizing it."

Once the customer opportunity has been created, the solutioning is where most people recognize innovation as taking place. Breakthrough innovation comes from finding radical solutions. This is where connecting a product or process from a totally different environment often leads to that "Aha!" moment. This is exciting! This is the application of new knowledge. Henry Ford's ideas for mass production came from seeing a meat processing factory and applying the factory's concepts to production of motor vehicles.

However, after this second stage we only have a *concept*. We now have to develop a working solution. This is where the developers take over. The third stage in the process of making the solution work is where so many organizations lose momentum and lose the advantage they gained in stages one and two. Speed to market is essential. Discipline becomes vital. This is where Thomas A. Edison's saying "Genius is one percent inspiration and 99 percent perspiration" is applied.

Finally getting to market is where far too many stumble. The production, service delivery, and sales people need to have been involved in the earlier stages if you want a long-term and continuous innovation process. They now have to "run for the line." Every advantage we can give them is essential. Time and advance notice are their biggest advantages of all. Too many companies wrap their development activity in excessive secrecy, and the sales and operations people are then presented with a challenge of which they had no previous knowledge.

The activities that need to be performed in an innovation process are:

1. Find the market need and opportunity
2. Identify potential solutions for the need
3. Select the solution
4. Develop the solution to be user-friendly
5. Take the solution to market

The organizational functions that perform these activities typically are:

1. Marketing and research
2. Design and quality

3. Leadership
4. Engineering and development
5. Operations and sales

Additionally, to innovate successfully an organization needs two modes of operation in its innovation process: "staying loose" to create and conceptualize and "hanging tight" to develop the concept and get to market. See Figure 1.2.

I will explain how to overcome this apparent paradox in the subsequent chapters.

CREATIVITY AND INNOVATION

People frequently confuse *creativity* and *innovation*. Creativity is only one of the attributes required for successful innovation. Creativity is a way of accessing new knowledge. It is also a way of retrieving existing knowledge that we may have forgotten, which we call *subconscious knowledge*. Creativity and innovation are different. Creativity is what initiates the innovation process, and people frequently think this is the whole process.

I was a member of the "Terms and Definitions" working group that wrote ISO 9000. That experience developed my skill with terminology.

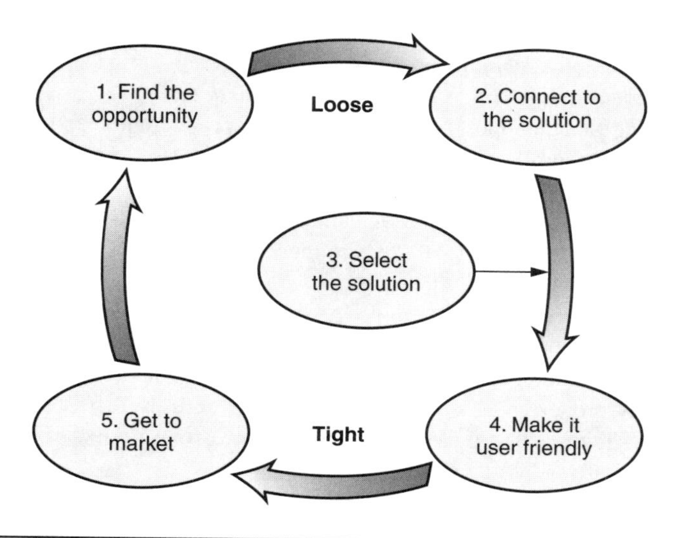

Figure 1.2 The innovation process—a paradox.

No definition is absolute, and most definitions in the English language have their roots in Latin, French, German, or Danish. When we look at "innovation" the word "nova" or "new" leaps out at us. On the other hand "creativity" is clearly about "creation"; this is about "building" or "developing" and is less about the final result.

You have seen that creating the opportunity and the potential solution are only the early stages of the process. Making the solution user-friendly and getting it to market are the later and tougher stages in the innovation process.

As a result of what I have explained we have a variety of definitions for innovation. The word "new" is fundamental. What becomes evident is that the fuel of the innovation process is "new knowledge." The final outcome can be a product or service, or it can be more esoteric and just be called "value." This gives us these potential definitions for innovation:

- The conversion of new knowledge into new products and services
- The conversion of new knowledge into new value
- The creation and introduction of a new product with value

However, an emerging understanding of innovation is that true innovation will cause the user of the innovation to "do things differently." It is not the product but the effect of the product or service that matters. Innovation is different from improvement in that it causes a different behavior in the user.

Innovation involves the successful implementation of creative ideas. Creativity is about *coming up with ideas* while innovation is about both *creating ideas* and *putting those ideas into practice.*

Innovation can apply to either a process or product and can be internal or external to the organization. Innovation is now also seen as a management process in an organization.

This means that an innovation can be a change in your business model or business processes or it can be a change in what you deliver to the customer, whether product or service.

The other challenge is to define whether innovations are "breakthrough," "disruptive," or "radical." Given the change in behavior that is implied, radical innovation can be described as a new product or service that causes an activity to be carried out in an entirely different manner. Research suggests that only about 15 percent, or one in six, of so-called innovations are radical.

In Chapter 2 we'll look a little deeper at this thing called "knowledge" that is so fundamental for innovation.

However, let me now draw your attention to the first of the Browser's Briefings at the end of this chapter. It summarizes the key points from Chapter 1.

BROWSER'S BRIEFING CHAPTER 1

- Innovation is about developing the products and services that the market needs tomorrow.
- Business has focused on improving the delivery of today's products and services and not on the development of new markets and products.
- Quality management has frequently become internally focused.
- If your product or service has become outdated, effort to improve efficiency is wasted.
- Good innovation strategy finds a new market, which the competition later sees no value in entering.
- Open market innovation recognizes that there is more knowledge "outside the box."
- Accessing knowledge outside the box involves sharing knowledge, and many have difficulty with this culture change.
- Innovators must recognize mega trends such as globalization, information technology, global warming, and aging populations.
- Finding a "cool new idea" has no value unless the problem that the idea solves has significance.
- New ideas will not be adopted unless they are easy to adopt.
- New ideas are most frequently the result of collective knowledge and do not come from a "lone genius."
- Innovation is driven by the need of convenience.
- Innovation follows a defined process.
- Creativity is only part of the innovation process.

2
The Need for Knowledge

The only competitive edge an organization has is the ability to learn faster than the competition.

—Arie de Guess in Peter Senge's book *The Fifth Discipline*

The twin towns of Kitchener and Waterloo celebrate the month of October with a Bavarian Oktoberfest, and October in Ontario is a beautiful time of the year. The leaves are turning and the colors are magnificent. The Kitchener-Waterloo section of ASQ invited me to speak at their October meeting and give my keynote speech "You Too Can Innovate," which I had given at the ASQ World Conference. The early evening drive from my home in Kilbride was an absolute pleasure. The colors of the trees brought out the artist in me. I always enjoy the drive to Waterloo and the town is also home to one of my favorite clients, Research In Motion (RIM), the inventors of the BlackBerry.

The evening was a pleasure and it was good to reconnect with old friends. The audience enjoyed doing the innovator self-assessment, which is part of the speech and is described in Chapter 3. At the end of the evening the section gave me a gift of the book *Leonardo da Vinci* by Frank Zollner.[1]

Reading the book the following weekend reminded me how at the time of the Renaissance we had not separated art and science. Leonardo was a writer, an artist, and an engineer. He is someone with whom I have a great affinity. As I read the book I marveled at his understanding of the laws of physics and at the same time his magical ability to work with color.

If you saw the movie *The Da Vinci Code,* you will be familiar with the balance that he brings to his painting of *The Last Supper.* Most of us also marvel at the subtle smile of the Mona Lisa, which is captured in her lips and eyes. In another of his works, *Madonna and Child,* the colors are also subtle but very well balanced. The words *subtle* and *balance* keep appearing.

Many people think the artist and the engineer are poles apart. One is loose and creative, the other is tight and analytical. The engineer takes an activity that was previously shrouded in mystery and through their analytical and mathematical skills breaks it down into logical steps. On the other hand the engineer is frequently critical of how the artist creates ideas out of "thin air" because these ideas are often intuitive and therefore do not appear to be substantiated.

Over the last century our educational system has increasingly separated these two views of life. It is convenient for the educator. It is not good for society. The fracture this has caused is also at the root of many of today's organizational roadblocks to innovation. Creative and analytical people are separated during their education and frequently remain separated during their subsequent work life. For successful innovation we have to reconnect the minds of the creator and the analyst. The left and right brains have equally important roles in the innovation process. The work of Leonardo da Vinci shows the importance of this balance. The challenge of the innovator is to create new knowledge, and this requires the left and right brains of both the individual and the organization to be truly connected.

LEVERAGING COLLECTIVE KNOWLEDGE

Information and knowledge must flow throughout an organization before it can become innovative. For information and knowledge to flow, the people in the organization must be linked.

A couple of years ago my mother experienced a fall and cut her leg. The hospital where she was treated, ironically named Good Hope Hospital, exacerbated her illness because the flow of information between people was so poor as to make the hospital organizationally incompetent. The individual staff and physicians were technically outstanding, but the hospital did not operate as a system. A primary problem was the inability of the organization to interface with external organizations. My mother contracted MRSA, which results from poor cleaning services. An organization is ultimately responsible for the results of all its subcontractors. This organizational incompetence showed up again in the inability of the hospital to interface with Social Services. A hospital that was supposedly short of beds took a full week to discharge my mother after she was ready to return home.

If you are not organizationally competent, you can't be organizationally innovative. You may have many separate individuals "innovating," but until they come together collectively you will not beat the competition.

An effective management system enables organizational innovation, and this is achieved through the flow of knowledge *between* people. To achieve rapid flow of information and knowledge and become a competent organization, the organization's processes, its people, and its technology need to come together as a system. This brings together the "left and right brain" of the organization.

Too often businesses try to cover their management system deficiencies by trying to dazzle the customer with a wonderful new product or service in the hope that they will be forgiven their previous sins. However, this type of innovation is only "skin deep." This is a short-term strategy, and their customers will find them out when they fail to deliver. They may be creative but they are not innovative.

So the question is, "How do we develop information flow in order to create organizational knowledge and apply this knowledge to become an innovative organization?" I am going to develop your understanding of how to create new knowledge and also how to manage new knowledge. I am going to do this at an individual and also at an organizational level.

INDIVIDUAL AND ORGANIZATIONAL LEARNING

At an individual level we gain information and we process this information to create knowledge. We call this process *learning.* Many individuals do not progress beyond learning. They become "sources of knowledge." See Figure 2.1.

Successful individuals apply their knowledge in a way that creates value for themselves and others. Their knowledge flows out from them in a way that "makes a difference." They are competent and are strongly positioned to innovate.

You can read as many books on golf as you wish, but until you have been on a golf course, you can't play golf. Talk to any golfer and they will also tell you that you can't be competent until you have played in a competition. Once you are a competent golfer you can innovate and try new ideas that will improve your game.

Individual and organizational innovation are similar. Organizations must also be competent to be competitive. To achieve the necessary flow of information and knowledge to become a learning organization, the organization's processes need to come together as a management system. You

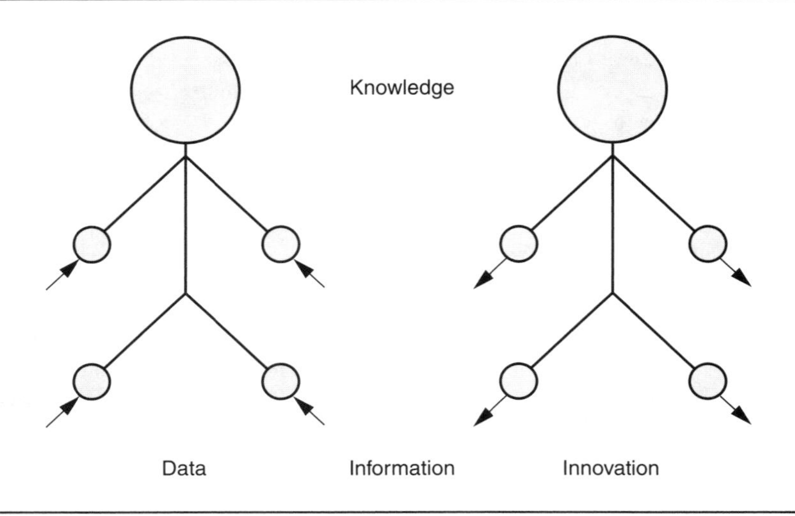

Figure 2.1 New knowledge leads to innovation.

probably viewed Figure 2.1 as a stick man, showing the learning process for an individual. It could also be viewed as a process map showing the learning process of an organization, with information flowing between processes.

When information flows between processes, organizational innovation is achieved through the application of organizational knowledge. To quote a good, and sadly departed, friend of mine, Herve Mignot from Paris, "The organization's head is connected to its feet."[2]

INFORMATION FLOW

For information to flow quickly we need the right organizational structure. The traditional organization structure inhibits information flow (see Figure 2.2).

In the past we have attempted to address the problem of information flow by viewing activities from a process approach and by process mapping. I think most implementers of quality management have process-mapped at some time. Figure 2.3 shows a typical process map example.

Process mapping is a technique adopted from the chemical industry. It has a lot of benefits but it treats your organization as if it were a chemical plant with processes represented by tanks connected by pipes. It only addresses the process aspects of your organization. If you have process mapped using the "swim lane" technique, which identifies people's

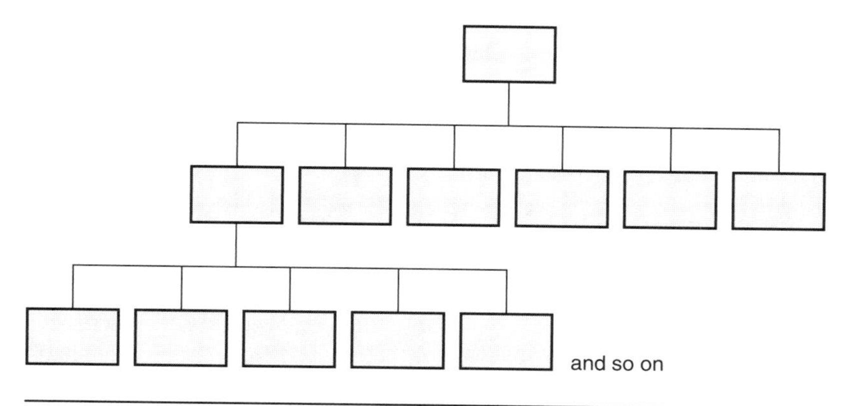

Figure 2.2 Traditional structure.

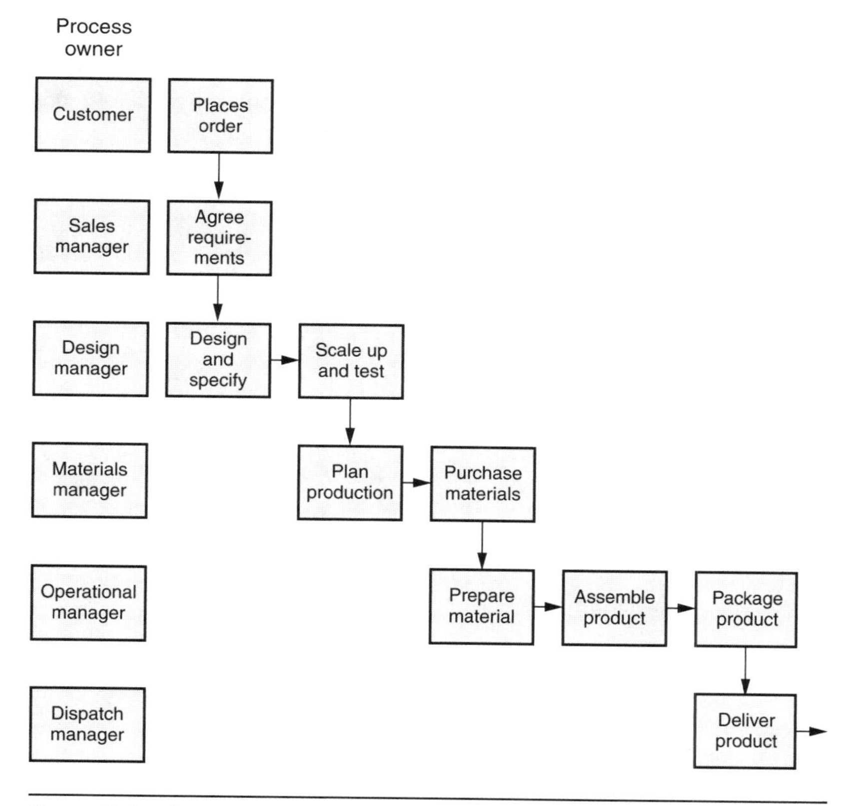

Figure 2.3 A process map.

responsibilities, then this gives you not only your process network but also the beginning of your "people network."

In the last 10 years we have moved to systems thinking as we have process-mapped our product flow. However, we have rarely mapped the flow of information between people.

NETWORK THEORY

This information flow between people is often referred to as the "network" of the organization, and in the past this network has relied far too much on random communication. An organization is a complex system. Network theory addresses the flow of information in complex systems. If we learn about the information flow in our network, it can enable us to manage our organizational knowledge.

The last five years have seen a tremendous growth in the understanding of network theory. You know that every organization has its own people network and how information flows between them. You need to understand the structure of your organization from a network perspective (see Figure 2.4). In a networked organization, as it grows, people stay closer.

This line of thinking was developed by Albert-László Barabási in his excellent book *Linked*.[3] I picked this book up in, of all places, the gallery bookstore after visiting my client Deborah O'Leary at the Art Gallery of

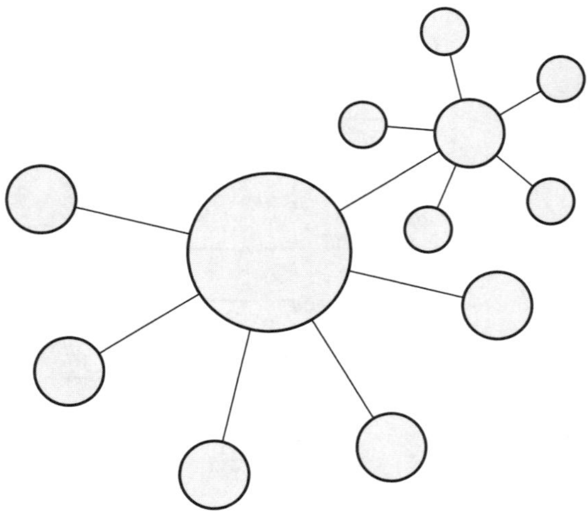

Figure 2.4 Organizational networks.

Ontario. Barabási develops network theory by comparing social, technical, and biological networks. He shows how we originally thought all networks were random, rather like a highway system with a small number of links to each of its nodes.

The key points in a network are its *hubs* and *nodes*. In the people network of an organization, these are its managers, its meetings, and social activities, whether formal or informal. In a successful organization formal and informal networks are the same; it is an organism, like a human body.

The word *random* is really saying "we don't understand." New thinking has developed the concept of the *scale free* network, rather like an air traffic network. Nodes (local airports) have a few links whereas hubs are highly connected. Take out a hub (major airport) and the network can be seriously damaged. Think about your own organization and how this applies.

The reason organizations experience so much difficulty with communication is because they do not understand their network structure. This structure is not the organizational chart and it is not even the process map. It is the links between the people or what we refer to casually as our *network*. You need to learn the primary information flows between the people in your organization. You can then understand how people in your organization gain knowledge and you can then manage your knowledge. Tools such as *social network analysis* enable you to do this.

Once you understand your information flow you can start to grow your organizational knowledge, and that knowledge creates new products and services.

KNOWLEDGE MANAGEMENT

Innovation is about the conversion of new knowledge into new products and services, and so an innovative culture has to be a learning culture. We are acquiring new knowledge all the time, and we store this knowledge in our minds. An innovative organization must be agile, so we must have the courage to allow knowledge to be stored in the minds of our people and must not be obsessive about documenting knowledge.

The advances in information technology make this documenting very tempting. However, documentation seriously reduces the agility of an organization, and we can only document a fraction of the knowledge in people's minds.

Documented knowledge is referred to as *explicit* knowledge and comprises less than 20 percent of our available knowledge. The knowledge in our minds is *tacit* knowledge. There is a whole raft of untapped knowledge stored in our tacit and subconscious minds (see Figure 2.5).

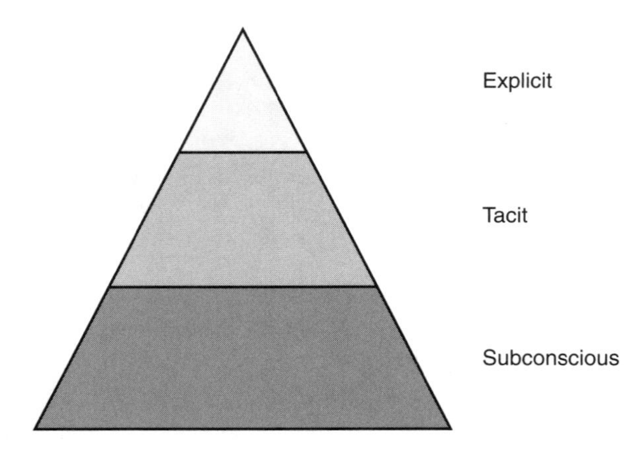

Figure 2.5 Releasing knowledge.

Knowledge is now being recognized as the fourth business resource alongside time, money, and people. Knowledge management (KM) is a natural outcome of the work of the last 25 years that has focused on quality and process management and which developed the ideas of learning and continual improvement. Continual improvement simply asks the question, "How do we do it right the second time?"[3] Innovation asks, "How can we radically change the way we 'do it' and also radically change what we supply to the market?"

To innovate we must grow our knowledge and develop new ideas from our knowledge. KM provides the platform for us to innovate.

HOW WE GROW KNOWLEDGE

Once you understand your information flow you can start to grow your organizational knowledge. However, KM is more than just the flow and storage and retrieval of information. It is also about learning and growing the knowledge of the organization and using that knowledge to create new products and services. Many think this is an IT issue. To say that KM is an IT issue is to say the minds of our people contain no knowledge. Europeans see KM as very much a people issue. This is a question of balance. Balancing people and process and technology is the key to success.

Remember the types of knowledge in an organization. The knowledge inside the individual is referred to as *tacit knowledge*. The other type is the knowledge that has been documented, called *explicit knowledge*.

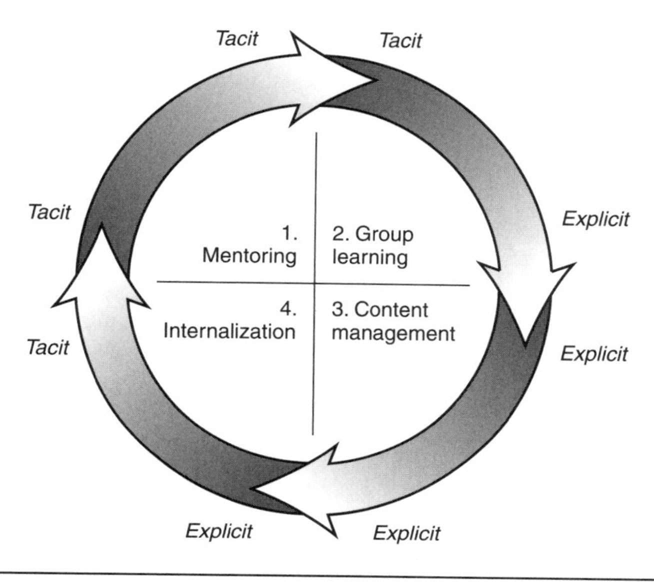

Figure 2.6 The knowledge management cycle.

The *knowledge management cycle* created by Nonaka and Takeuchi (see Figure 2.6) shows the importance of the rapid transfer of tacit and explicit knowledge between people.[4] It has four stages. It is a simple model that groups tacit and subconscious knowledge together.

Stage 1

For centuries the first method of learning has been analogous to the "sorcerer and the apprentice." The secret of success in your life was to find a good mentor. This process involved the transfer of tacit knowledge between two people. From when you were taught how to ride a bike to having lessons from a golf pro, you learned the hidden mysteries, you were "personally coached."

Stage 2

As organizations have grown, the ratio of available mentors has declined, so group learning has become important. One of the earliest recorded examples of group learning is Aristotle teaching Alexander the Great and his contemporaries. This is how Alexander was able to develop his elite officer corps. A key feature necessary for this process to succeed is *dialogue.*

Knowledge is also documented during this process and moves from tacit (stage 1) to becoming explicit. This stage of group learning is also where we introduce one of the best-known KM tools, *communities of practice,* which then morphs into *communities of innovation.* I will talk about this more in Chapter 12.

Stage 3

As the bodies of knowledge in the world have grown exponentially in the last decade it has become necessary to store explicit knowledge in such a way that it can be easily accessed. The ancient profession of librarian has taken on a whole new meaning. This *content management* is a subset of knowledge management. This is where organizations have wasted mountains of money and built mountains of frustration through badly implemented IT systems. The move to overdependence on explicit (documented) knowledge has reduced the ability of organizations to innovate.

Stage 4

Explicit knowledge must be stored in a way that will enable it to create value. The core problem is not the IT system but the method by which tacit knowledge is recorded and the method by which explicit knowledge is stored. In the past, both have been very badly executed. The objective in this last stage is for people to be able to retrieve explicit knowledge and make it become tacit knowledge. This rarely happens successfully.

So to innovate successfully you need to develop linkage of processes, people, and technology, and these are primary attributes of an effective management system. The belief that technology will solve all information issues ignores the important transfer of information and decision making that occur during meetings, whether in groups or one on one.

To emphasize how the flow of knowledge and information between people takes place during meetings and in communities of innovation, let's look at a comparison of documents and meetings (see Figure 2.7). (*Bandwidth* denotes the amount of information transferred in a given time.)[5]

Taking this a step further, the book *The Wisdom of Crowds* by James Surowiecki shows how breakthroughs come not from the "genius" and how collective knowledge is so much more powerful.[6] If you go to a racecourse, the bookmakers do not set the odds based on their own knowledge and experience even though it is probably far more extensive than the race goers. They rely on the "wisdom of crowds." I explain this 'wisdom' more fully later in Chapter 12, "Skunk Works" and Chapter 14, "Networking."

	Interactivity	Bandwidth	Reusability
Document	Nil	Low	High
Meetings	Very high	Very high	Low

Figure 2.7 Transfer of tacit knowledge between people.

The challenge for an innovative organization is to release tacit and subconscious knowledge and convert it into new products and services.

The technique for doing this is *networking,* which recognizes that if you combine the tacit and subconscious knowledge of two or more people you have a powerful combination. Collective knowledge is far more powerful than the knowledge in any one individual or in any manual.

Tapping into this collective knowledge requires us to radically change many of our behaviors. We have traditionally veiled new ideas in a cloak of secrecy, we have traditionally managed our time so there is not a single wasted moment in our day, and we have increasingly focused our improvement efforts on our processes and not our products. Finally, embracing failure is something we all struggle with. If you want to innovate, all of this must change. I will talk more about this in Chapter 11, "Culture." One more thought on releasing knowledge follows.

TAKING A BATH

Rugby is an intense game. As I write, the Rugby World Cup is being played out in France. Countries from the Asian Pacific, Europe, Africa, and the Americas are competing. If you watch a game of rugby you will see both the physical and mental intensity that goes into it.

What you don't see is the huge amount of preparation beforehand, which is not just physical, but also involves an enormous sharing of knowledge between the people on a team. I played rugby for many years. Like many players the only thing that stopped me, apart from the inevitable sports injuries, was the inability to give time to training.

What you also don't see in a game of rugby is the time invested before and after the game. There is a popular image of rugby players having a long hot bath then drinking lots of beer after the game. Yes, these things happen, but you need to peel back a layer and understand the sharing of experiences, the reliving of the game, and the learning that is taking place.

As with Archimedes, having a hot bath and relaxing is one of the best ways of releasing knowledge. His "Eureka!" story is of course one of the great stories in the history of physics. We know equally well that relaxing over a glass of wine, beer, or a hot drink releases many thoughts that have been locked in our subconscious.

In the modern world of business we don't take enough time to do these things. New knowledge is the fuel of innovation, and that knowledge comes from you and the people you interact with.

So the next question is, "What is your own best role in this process?"

BROWSER'S BRIEFING CHAPTER 2

- Separating the left and right brain reduces our ability to innovate.
- An innovative organization enables information to flow freely between its people and also enables information to flow freely to external organizations.
- An organization must have an effective quality management system to innovate successfully.
- A learning organization becomes a competent organization.
- The traditional hierarchical organization structure inhibits learning while the process approach improves learning.
- A networked organization is the most effective at transmitting information and knowledge.
- Social, technical, and biological networks function in the same manner.
- For successful innovation, the internal and external networks of the organization must be developed.
- Knowledge management (KM) is the platform from which innovation is developed.

Continued

Continued

- Documented or explicit knowledge accounts for only 20 percent of the knowledge in an organization.
- An innovative organization enables the rapid transfer of the knowledge in people's minds (tacit knowledge) by using its network.
- We have to allow time for the release of tacit knowledge.
- Communities of innovation are used to initiate innovation.

3
The Roles

We all have ability. The difference is how we use it.

—Stevie Wonder

When my daughter reached the age of 15 she started to talk about career options. She was very clear on her choice; she wanted to become a beautician. Well, at the age of 15 there are few girls who would not like to become a beautician. It's a bit like boys at the age of 15 wanting to become a rock star. We thought about this, and a friend of mine, Bill Rowney, told me he had taken his son to vocational guidance counseling and how happy they had been with the outcome.

So that's what we did, and the counsel we received was that because of her spatial ability and interpersonal skills Rachel was the perfect fit for her chosen career. We sent her to a specialist college and she qualified as a member of the Society of Aestheticians. She specialized in skin care and after a very successful and happy career as a beautician she went on to be a spa manager. I then think of my own career and how receiving a chemistry set as a child sent me down the road of being a chemical engineer. However, my skill sets emerged in the field of writing, speaking, and training, which is where I am now, but it took some time to get there.

So many of us struggle to find our role in life and it is so rewarding to see someone find their perfect fit. I am reminded of the words of a friend of mine who is a counselor. Joe said to me once, "Do what you do well." Simple words, but so often we struggle in a job or career where we arrived through a set of circumstances that were not of our making. We may do well in the job, but we may not be happy in the job. Too often we see another job and it looks attractive, so we are drawn to that type of job but we do not have the aptitude for it.

Innovation is the same. Many people say, "I'm not creative so I can't contribute to innovation." In truth the toughest job in innovation is getting the new product or service to market, and this requires people who are far more results oriented. Other people say, "I'm not an ideas person," but their skill may be in taking an original idea and refining it.

KNOWING YOUR BEST CONTRIBUTION

Knowing where you will make your best contribution to innovation is essential. Figure 3.1 shows a chart that helps you assess where you will make your best contribution. A word of caution: if you score highest in, say, column four, which is the *developer,* (by the way, that is my highest score) you can still contribute in other areas. In fact you will see that it is essential to maintain a mix of all types of people as you move through the innovation process. On the other hand, it would be fatal to perform the development stage of the process without a majority of people who scored highest in that category.

One other word of caution: don't look at the column with your lowest score and rush to try and improve that score. Remember, "Do what you do well!"

The key roles in the innovative organization and in the innovation process match the stages in the process (see Chapter 4). If you conduct this self-assessment it will show you where you will make your best contribution. Do you make the best contribution to innovation through:

- Generating opportunities (creators)
- Linking those ideas to solutions (connectors)
- Turning ideas into practical solutions (developers)
- Implementing solutions and getting things done (doers)

So what will be the best contribution you can make to innovation? Will it be through generating ideas, linking those ideas together, turning ideas into practical solutions, or implementing solutions and getting things done? We all have a role to play. Score yourself on Figure 3.1.

1. Look at the first row (I like to find answers, I like to finish, I like to explore, and so on). Score four points for the phrase that *best* describes you. Then on each row below, score four points for the phrase best describing you.

1	I like to find answers		I like to finish		I like to explore		I like things to work	
2	I need to understand an issue		I make things work		I see both sides of an issue		There has to be a right answer	
3	Don't tell me what to do		Give me facts not theory		I create choices		I like to analyze data	
4	I am open-minded		I convince people		I have lots of ideas		I find a weakness	
5	I "connect the dots"		I get things done		I like possibilities		I bring things "down to earth"	
6	A concept must be sound		I like "energy"		I don't fuss with details		I like precision	
7	I do not like confusion		I avoid theory		I avoid decisions		I do not like failure	
8	I think things through		I take risks		I like to hear about problems		I focus	
9	I like solutions		I like end results		I like opportunities		I like simplification	
10	I want to own the problem		I find a way that works		I like the big picture		I am thorough	
11	I like to define the problem		I want agreement		I find out the facts		I plan	
12	I want ideas		I want to try new things		I want space		I want structure	
	Total		**Total**		**Total**		**Total**	

Figure 3.1 Creator, connector, developer, doer?

2. Go back to the first row and, working down the rows again, score one point for the phrase in each row that *least* describes you.

3. Finally, score three or two on each row for each of the remaining phrases—three points for the choice that is more like you and two points for the choice that is less like you (you must have a four, a one, a three, and a two in each row)

4. Total your columns.

Row 1—you're a connector, row 2—a doer, row 3—a creator, and row 4—a developer.

Which column contains your highest score? Column 3, the gold-diggers, typically find the opportunity and open it up. If you do this you are likely to be an artist, marketer, or researcher. You are practical and you learn from practical experience.

Column 1, the connectors, are the "green thumbs" who nurture the seed of the idea and find the answer. They are the design, R&D, and strategic planning people. They are a rare breed and learn by thinking.

Column 4, the developers, also learn by thinking but make the idea work. They are engineers, systems developers, and accountants.

Column 2, the doers, finally take it to market and get the job finished. They are project managers, salespeople, or folks from production. Like the gold-diggers, they are practical people and like the gold-diggers they learn from practical experience. One other point—doers and connectors are opposites.

WHAT DO THE SCORES MEAN?

In Chapter 12 I will explain how you form your community of innovation, and you will see that diversity is important so that ideas will be challenged. Doers and connectors are opposites, while developers think creators are not focused. On the other hand, creators think developers don't see the big picture. You can see a tension building, but this is important.

When people have completed the scoring they often ask, "I have two scores that are almost identical, what does that mean?" I can't address all scenarios—that would require one-on-one discussion—but there are a couple of common situations. First, it is clear that you can contribute in both areas of the process. However, if your scores for connector and developer were similar, it means you like to think things through, whereas if your scores are the same for developer and doer, you work best on a deadline and you thrive on time management and efficiency.

You know from your own experience that some people are good at thinking things through while others learn from experience. In truth our minds are more complex than this and we are all a mix of both. For my own part I am labeled as a *thinker* and yet I know that I have to do something practically for the experience to lodge in my mind.

Let's take a closer look at each of the roles.

CREATOR

Creators learn from experience. If you are a creator, you see something happen and the event impacts you. Your mind races on to see other opportunities related to the event you have witnessed. You see unfulfilled needs. However, your mind does not like boundaries and you will flow from one opportunity to another. You are an observer, but you need to capture your thoughts before they move on. The skill of note taking is something you should develop, though this may be alien to you. I talk about this further in Chapter 13, "The Competent Innovator." Because you don't like boundaries, you prefer your ideas to be "gray" at the edge. You don't like people asking you to define a problem and you don't like people asking you to make decisions. In fact, you like to keep on generating more and more choices. You operate best in a loose mode, free of boundaries. You may have disliked those school exams where there was a "right answer" but probably enjoyed those where you could just flow with your thoughts. You love possibilities. You are an exciting person to work with, but beware of conflict with developers. They are highly focused people and they will accuse you of "lacking focus." That's okay, that's not your job. My wife is a creator and I am a developer and we work well together if we remember that our aptitudes lie in different areas. As a creator you also need the opportunity to explore in order to succeed. I talk more about exploration in Chapter 13.

As a creator you will be most comfortable in an artistic or research environment. This means graphic arts, market research, or an activity with a lot of human interaction. Ask yourself if you are doing what you do well.

CONNECTOR

The creator's best friend is probably the connector. If you scored equally on both you are interchangeable, which speaks to the fact that both types work best in a loose mode. Although the name has changed, both people operate in a creative mode. However, where creators are the problem finders, connectors are the problem solvers.

The linkage between creators and connectors is important. Creators do not like to put a definition on a problem whereas connectors do. In a large population you will find that true connectors are in short supply. That is why people providing training in problem solving have had open season during the last 20 years. Although generally in short supply, in the world of quality management there can often be an above average number of connectors.

The quality management profession attracts problem solvers. I mean *true* problem solvers. Most people jump to a solution and implement. The danger in the world of the innovator is that the creator sees an opportunity and we implement the first solution that springs to mind. Connectors don't do that. Importantly, they define the problem the creators have uncovered and, even more importantly, "connect" to solutions. What I mean by this is connectors will take solutions from a different context and put them into the context of the current problem. A good example of this was when Henry Ford was seeking the solution to his problem of mass production and saw a meat processing plant where the carcasses were being moved by hanging them on hooks. This "connecting" happens to all of us at some time and it often happens so quickly we think it is "magic." In truth, connectors can have worked on a problem for a long time before they find a solution. The danger therefore is stopping at this one solution, so relating back to creators is important. Remember, they like lots of choices. Connectors love problem solving and so they need a diet of problems. If you are a connector, compared with creators you are far more of a thinker. Because you are a thinker you will really want to understand the problem and you will address solutions from a conceptual rather than a practical level. However, you will want your concept to be sound.

Because you are a thinker you may not be involved in implementing the solutions. Don't be insulted, that is not your job. Linus Pauling the Nobel Prize winner said the best way to have a good idea is to have lots of ideas. This applies to both creators and connectors.

SELECTING THE SOLUTION

Selecting the preferred solution is something I will discuss later. This is the job of the strategic planners. The connectors need to provide them with choices and those choices need supporting data on risk, ROI, and practicality.

Many people think that once you reach this point, innovation is finished. What I have described so far is not innovation, it is creativity. Creativity is a subset of innovation.

Innovation is about converting new knowledge into new products and new services. So far we have only created new knowledge. We do not yet have the new product or service. Now the game changes and we go from "loose" to "tight." We have to make the solution practical and we have to get it into the hands of the user.

DEVELOPERS

The developer is far more focused than the connector and the creator. If you are this person you will work best in a defined or project-driven environment. You want a specific problem to work on and that problem must not be ambiguous. The many ideas from the connector must have been distilled into two or at the most three choices to enable you to focus. You are similar to the connectors in that you are both thinkers and so you need to take time to understand the concept they have developed. Your strength is that you turn abstract ideas into solutions that work. You are good at data analysis and so can pinpoint weaknesses in a product or a process and move on to the best solution.

Time is of the essence at this stage in the innovation process, and you might find this problematic because you want your solution to be precise and unambiguous. You hate being taken off the job you are working on before it is complete and being asked to work on something else. You will miss lunch or dinner if you have in your mind a completion point for something you are working on.

This is the attribute where I have the highest score, and I envy those people who can leave off a task that is incomplete, relax and enjoy lunch, and then come back to the job later. Taking time to relax is a skill that developers have to work on. This is when developers can have some of their best insights. Remember that Archimedes found his answer to specific gravity when he took a bath. A lunch "out of the box"—and I don't mean a "boxed lunch," I mean out of the building—can often be your best friend when you are hitting the wall and not finding answers.

Because you are focused you can find yourself in conflict with the creators, who like to see the big picture, and you can also forget to include the potential users of your solution. Don't forget the "alpha testing" that is fundamental for the developer. You are probably a good planner so use that skill to create your project timeline and also to involve the implementers or doers who come next.

DOERS

The doers, like the developers, work best in a project-driven environment and need the tight mode of operation. You may think that if you scored highest here you have no job in the world of innovation. In truth, your job is the toughest of all. You need to have been involved with the developers so you know what is coming down the pike. You are similar in that you work best in a structured environment but you are different in that developers are thinkers whereas you like to "do," as your title suggests. Be aware of this and respect your differences. You may be surprised to find that you have strong similarities to the creators in that you are both practical people. You both enjoy new situations and find them stimulating. You can be impatient and, unlike the developers, you will keep "breaking things," trying different solutions until you find something that works. Use the thinking strengths of the developer to help you here. You like getting your feet wet and your hands dirty. You feel you are getting something done.

You are likely to be in operations or sales, if you are in your natural habitat, because you like to deliver. Because of your practical approach to life, if you are in operations, you find to your surprise that you mix easily with the sales and marketing people but you are less at ease with those R&D folks. If this is the case, be conscious of it and learn to work with those "thinkers."

Your challenge is to get the solution in the hands of the customer. If you're an operations person this means working with the developers to eliminate production or service problems. If you are a sales person you need to understand the ideas of the creators in marketing who originally saw the opportunity.

LEARN TO WORK WITH OTHERS

You can see that whether you are a creator, connector, developer, or doer, this does not put you in a box. You need to work with people that have other attributes in order to do your own job. Understanding how those other people think and act will help you to work better with them. Use the assessment on your team and understand your colleagues.

These revelations may not be new to you but they are indicators of where and how you will make your best contribution to the multifaceted world of innovation.

For success you need a mix of people at each stage in the process. The tension is healthy!

So having identified our role in the organization, what is the process we must follow?

How do we bring this mix of people together to innovate?

BROWSER'S BRIEFING CHAPTER 3

- We all have a role in the innovation process.
- Creators generate opportunities.
- Connectors link opportunities to solutions.
- Developers make solutions practical.
- Doers implement solutions.
- A self-assessment can show you and your colleagues where you will make your best contributions to innovation.
- Creators and doers are practical.
- Connectors and developers are thinkers.
- It is essential to have a mix of different attributes at each stage in the innovation process.
- The mix can create tension, but this is healthy.

4

The Process and the System

The best way to get a good idea is to get lots of ideas.

—Linus Pauling

Some friends from England, Phil and Liz, wrote to us one year and suggested that for the summer vacation we do a "house swap." It reminded my wife and myself of the movie *The Holiday* in which Cameron Diaz house-swapped with a person from England, and we smiled. Then we thought, what a great idea! They would live in our house in the country in Canada for two weeks and we would live in their apartment by the sea in England.

We agreed on the dates with our friends, made the travel plans, and then about two months before the swap it was a case of "reality bites." Our two friends were very tidy in their habits but we were a couple who lived life "on the run" with the attendant trail of debris.

About two years previously we had borrowed a book on getting rid of clutter and that suddenly became our bible over the following weeks. We discovered things we didn't know we had. We dumped things we hadn't used for years and we transformed the house into a place where there was not only room to store things but we could also find them! It was a great feeling.

My garage was like the world of the mad inventor: old tools, bits and pieces of metal and wood, old garden furniture, and a host of items that I was going to repair some day when I could locate the tools I needed to do the repair. I emptied everything onto the driveway, hosed out the garage, and returned to the garage only the things I really needed. The rest I dumped.

My point? Your innovation process, even in its early stages, needs to be free of junk. Your ideas need to be carefully stored and well organized like any good garage.

The world of the engineer and the inventor is often like my garage before it was de-junked. We keep things because "they might come in useful." This applies to both documents and materials and components. After a while we don't know what we have and we can't find what we want.

This happens due to the false premise that technology drives innovation. It happens through failing to understand the innovation process. Too often the inventor creates something that "might be useful" and then spends forever looking for a problem that their solution might solve. They think innovation is a solution looking for a problem.

Occasionally this works, and of course these are the stories that make good newspaper copy. The classic story here is of course the Post-It note.

In the early 1970s, Art Fry was in search of a bookmark for his church hymnal. Fry's colleague at 3M, Spencer Silver, had developed a failed adhesive which peeled off. Fry used some of Silver's adhesive on the edge of a piece of paper. His hymnal problem was solved! Fry saw that his bookmark could have other uses and the Post-It note took off.

The other story, in some ways more impressive, is IBM's alphaWorks. IBM is a great innovator. Their "garage" was so full of great inventions that were not being used that the development staff was angry and frustrated. So they had a "garage sale." They put all their unused ideas on a Web site for free download for 90 days. Forty percent of their ideas were taken up and they now have a $2 billion a year business selling previously unused products.

These stories make good reading, but they are not good innovation stories.

The truth is that the great inventions are driven by need and not technology and that the need is not always recognized. Yes, the second stage of the innovation process may generate solutions that could work in a different context. Yes, there will be many failures before the final success. However, if you clutter your life with your failures, your innovation will grind to a halt. Learn from the failure and move on. A good maxim from the world of knowledge management is that you are unlikely to use explicit (documented) knowledge that is more than two years old. It will either have become tacit or have become out of date.

Innovation does not start with finding a home for a cool idea, it starts with a market need and seeing a market opportunity.

To innovate successfully you must always be thinking about tomorrow's customer and what their needs will be. Much of quality management has tended to focus on today's customer and address their immediate concerns. When quality management has been inwardly focused it has often caused us to neglect the future.

The driver for the innovation process is obscure, and that is why people sometimes believe there is no process. The driver is tomorrow's customer, and of course that person does not yet exist. Phil Crosby said "All work is a process," and yes, innovation is a process. However, it is not linear; it is a looping process. It also changes mode from loose to tight and back to loose.

The reason most people struggle with the innovation process is that it is unusual. To innovate successfully, an organization needs to manage a paradox: "stay loose" to create and conceptualize, and then "hang tight" to develop the concept and get to market. Loose means allowing freedom to think; tight means being fast and efficient. These are the stages in the innovation process and also point to the key roles in the process.

The idea of the lone genius is a common myth about innovation. Most innovations are the result of collective knowledge. You can see by comparing Figures 4.1 and 4.2 that the stages in the process match the roles in the process. People with different attributes are needed at each stage:

- Generating opportunities needs creators
- Linking those opportunities to solutions needs connectors
- Turning ideas into practical solutions requires developers
- Implementing solutions and getting things done is the job of doers.

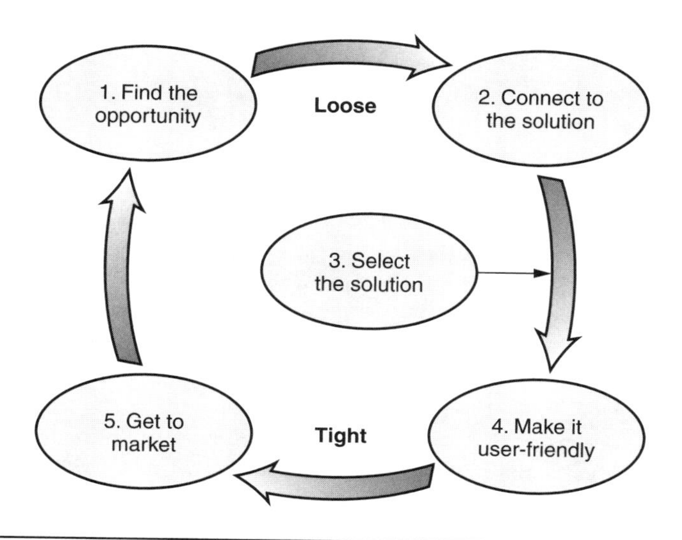

Figure 4.1 The stages of the innovation process.

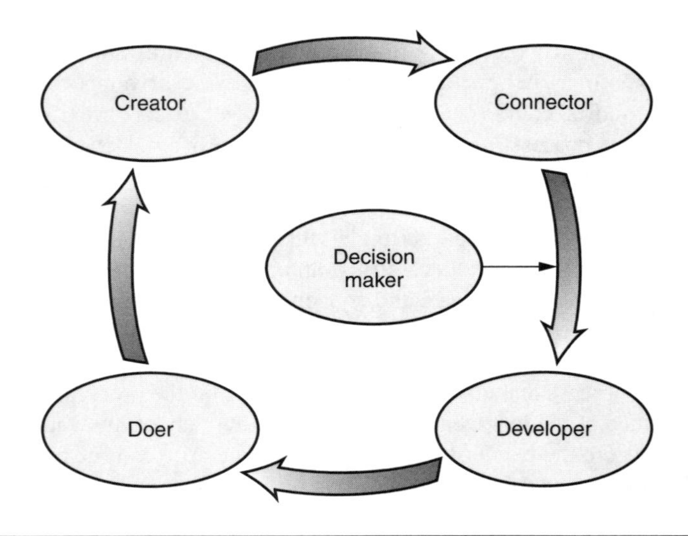

Figure 4.2 The innovation cycle—the roles.

If you conduct the self-assessment in the previous chapter it will show you which of these categories you fit into and the stage where you will make your best contribution to the innovation process.

THE OPPORTUNITY

The first step in the innovation process is to identify the opportunity. Importantly for the innovator, neither the customer nor the market may recognize that need or opportunity. Henry Ford said, "If I had asked my customers what they wanted they would have said 'faster horses'." Creative thinking is usually required at this first stage, and the creators are the primary influence at this stage. They are the "gold-diggers" who typically dig out the opportunity and open it up. These people operate best in a loose environment.

Typical examples of such opportunities include: IKEA found furniture stores overpopulated with aggressive sales staff and providing a confusion of choices; Southwest Airlines found that travelers were annoyed with bad meals and the large amount of leg room given to first class passengers; Yellow Tail Shiraz found that casual wine drinkers were intimidated by "antique" labels; and Cirque du Soleil found that the three-ring circus caused confusion and that there was increasing concern about the treatment of animals in circuses. I will talk more about these cases in Chapters 6 and 7.

But beware: people often jump from "create" to "do." People often rush to market with a half-baked idea. There are vital interim stages in the innovation process.

THE SOLUTION

Next, you find the idea that solves the problem. Once the customer opportunity has been created, then the solutioning happens. This is where most people recognize innovation as taking place. Breakthrough innovation comes from finding radical solutions. This is where connecting a product or process from a totally different environment often leads to that "Eureka!" moment. This is exciting! Innovation is the use of new knowledge to create new products and new services. This second stage is the application of that new knowledge. Again referring to Henry Ford, his ideas for mass production came from seeing a meat processing factory and applying its methods to production of motor vehicles. Connectors, like creators, need a loose mode of operation.

This will often mean removing what have been regarded in the past as sacred attributes of your offering. The companies I mentioned all had the courage to do this. They all took out what were previously considered sacred attributes in their respective industries.

THE TIPPING POINT

After the second stage we only have a concept. We then reach a tipping point where decisions have to be made. We make decisions on which is the best concept to pursue based on factors such as risk and behavior change. We then narrow our focus. This is also where our activity changes from the loose mode of the first two stages to a tight mode where the developers take over. Statistics show that out of 3000 ideas only one will come to fruition. Unfortunately, we often kill the best ideas with overaggressive requirements for ROI. Or we let a high-risk product through because we failed to assess the risk attached to a potential new product. We now have to develop a working solution.

DEVELOPMENT

The development stage should be fast and not secretive, and should involve the ultimate user. This is where you make the product functional and user-friendly, and eliminate the glitches. Good project monitoring and control are vital. This is where you move from loose to tight. This stage in the process is where so many organizations lose momentum and lose the advantage they gained in stages one and two. Speed to market is essential. Discipline becomes vital. This is where Edison's "Genius is one percent inspiration and ninety-nine percent perspiration" is applied. The developers make the idea better and make it work. They are engineers, systems developers, and accountants. Many companies go wrong here by staying in a loose mode of operation. It is vital to go tight.

EXECUTION

Finally, execution is where the operations and sales people take ownership, but remember, they must be involved in the previous stage so that there are no surprises here. Production problems are eliminated during development. The value proposition that the sales people will use is created during development! Getting to market is where far too many stumble. The production, service delivery, and sales people need to have been involved in the earlier stages if you want a long-term and continuous innovation process. They now have to "run for the line" so a tight mode continues to be vital. Every advantage we can give them is essential. Time is their biggest advantage of all. The doers execute and get the job finished.

We will look at these stages in more detail in Chapters 6 through 10. You need to know how good your organization is at each stage in this process. Assess your process at the end of each of the chapters!

I have given you a core process for innovation but it must sit within a management system if it is going to succeed.

THE QUALITY MANAGEMENT SYSTEM AND INNOVATION

Peter Senge's book *The Fifth Discipline* quotes the famous statement "The only competitive edge an organization has is the ability to learn faster than the competition."[1] I will extend this to say, "The only competitive edge an organization has is the ability to create and apply new knowledge faster than the competition." To achieve this flow of information and knowledge and become a learning organization, the organization's processes need to be networked to come together as a system.

In a successful organization the business processes link to form a management system and this enables the free flow of information, which in turn leads to the creation of new knowledge. Knowledge feeds innovation and innovation feeds success.

An effective quality management system is an essential platform for the innovator. This is how you execute effectively and protect yourself from the competition entering your market.

You can come up with the best new product in the world, but if you can't deliver, the competition will jump in and fill the vacuum of demand you have created.

An effective quality management system is lean and flexible. It engages the people in your organization and it produces nonfinancial data, which in turn produces knowledge you can use for innovation of both your processes and your products.

QUALITY MANAGEMENT PRINCIPLES

Quality management has a philosophy based on principles that come from the experience of businesspeople worldwide and have been documented in the ISO 9000:2005 standard, *Quality management systems—Fundamentals and vocabulary.*[2]

Looking at this philosophy from an innovator's perspective, the principles start by showing that a successful organization is *customer focused.* Innovation is driven by a focus on both today's and tomorrow's customer. The job of the *leadership* is to take this customer focus and set direction and create objectives for the organization. These objectives need to address the needs of today and also the needs of tomorrow in the form of new products and services.

The leadership must then create an environment in which the *people become involved* in achieving the objectives. People are the essence of the organization. They are fundamental to achieving its objectives and fundamental to innovation.

The most efficient way of utilizing an organization's resources is through *process management.* However, the processes in the organization need to come together as a *system.* An organization must seek *innovation*

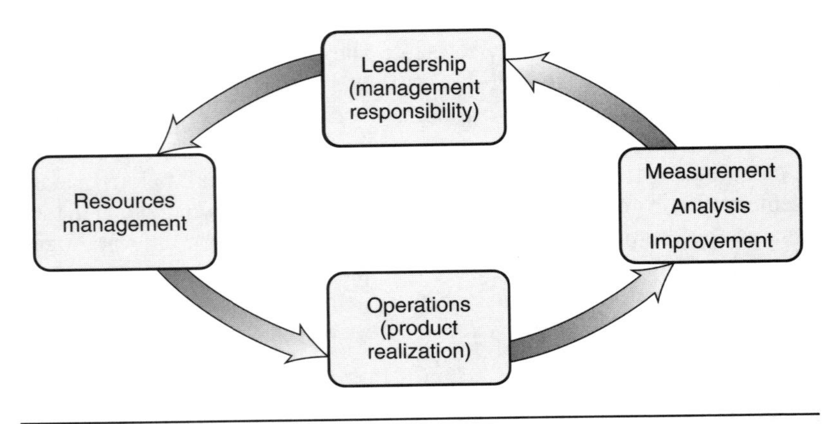

Figure 4.3 Create new knowledge from your management system.

of its products and services and processes. If you don't, your competition will overtake you. We drive this innovation through a *factual approach to decision making.*

Finally, *mutually beneficial supplier relationships* are what give the greatest value to the customer and to the organization. This concept recognizes that a business is not an island, but operates in a business ecosystem. Our suppliers are in fact a great source of knowledge, and open market innovation engages them as business partners.

ISO 9000 has taken these principles and given us a business model based on the plan–do–check–act cycle.

Knowledge is created at each stage of the cycle but emerges in the measurement and improvement stage. At this last stage we are collecting data, which we analyze.

Data are just numbers, but if we look for patterns in the data, when we find those patterns they give us *information. Knowledge* is information that we can act on. Innovation comes from the conversion of new knowledge into new products and services. This applies at both the individual and organizational level. Your management system must create organizational knowledge for you to be able to innovate.

An effective management system also creates *organizational competence,* and this is achieved through the flow of knowledge *between* people. Individual and organizational competence are similar.

Recall from Chapter 2 that this is the *learning* process and also remember the quote from Peter Senge's *The Fifth Discipline.*

THE INNOVATION MANAGEMENT SYSTEM

There are three levels to innovation management just as there are to quality management. They are *system, process,* and *product.* For *system* you can read *organization*; for product you can read *tangible* or *intangible.* Delivery of intangible product means the delivery of a service or of knowledge.

The IBM Global CEO Study[3] in their report *Expanding the Innovation Horizon* showed convincingly that organizations who innovated at a system or organizational level grew operating margin more rapidly than the competition. This is not surprising to people experienced in quality management and who know, to quote Deming, that 90 percent of the problems in a process are the result of the system in which the process operates. In turn, the way you get your product right is by focusing on the process that produces it.

An *innovation management system* (IMS) is fundamental to successful innovation. An innovation process is fundamental to the creation of new products and services.

As we move to integrated management systems, the IMS and quality management system (QMS) must clearly be linked. The QMS creates data, information, and knowledge, which are the feedstock for the IMS.

This doesn't mean you only innovate at one level; you must innovate at all three levels, system, process, and product. If you don't get your products, services, and markets right, the other stuff doesn't matter. Product innovators also outperform their competition on operating margin.

This brings us to the strategic decisions we must make as innovators.

BROWSER'S BRIEFING CHAPTER 4

- The innovation process must be kept tidy and organized.
- Innovation is driven by need and not technology.
- The early stages of the innovation process operate in a "loose" mode, the later stages in a "tight" mode.
- The opportunity stage is where we find a market need that is not satisfied.

Continued

Continued

- The market may not realize the need exists.
- The solution stage is where "connecting" to a different environment often reveals a radical solution.
- The best way to get a good idea is to get lots of ideas (solutions).
- A preferred solution is selected based on time, cost, and risk.
- The solution must then be made user-friendly to enable adoption.
- Executing the working solution is the toughest challenge of all.
- Only one out of 3000 ideas "makes it."
- An effective quality management system is essential to enable delivery at the final stage of innovation.
- The innovation process must integrate with the management system of the organization.

5

Innovation Strategy

If you see a bandwagon . . . it's too late.

—James Goldsmith

Business is about competing. Or is it? We are so focused on the competition that we miss new opportunities that are staring us right in the face.

My friends tell me I am too competitive, and in many ways this is because I was raised in an education system where unless you were first, either academically or in sport, then you had failed.

Competition was in my blood, but it seemed I was always trying to be first and not quite making it. However, competition is not always good. It can blind you to reality and to opportunities on your own doorstep.

Twenty years after leaving my school, King Edward VI School in Aston, Birmingham, the school had its 100th anniversary reunion at Penn's Hall in Sutton, Coldfield. It was early evening and I was at the bar with my friend Bob, ordering drinks for us and our wives. Nearby were a couple of guys a little younger than us and one looked at me and said, "Aren't you Peter Merrill?" I looked at him a little surprised, as I didn't recognize him and simply said yes. "Wow," he said, "you're Merrill, the runner! We used to think you were a god!" I was totally stunned. This was a new reality. After so many years as seeing myself as an average sportsman there was someone who saw me as a great competitor.

It made me realize that in our striving to compete we miss so much that is right there in front of us. Most businesses focus on beating their competition and they compete on small cost and quality improvements to existing products. They overlook the fact that customers are making major substitutions for their product by purchasing radical alternatives from other industries. Partygoers stopped buying wine and instead bought vodka and cranberry juice. People stopped going to the circus and started going to Disney. These are simple substitutions. Many substitutions are more extreme.

In business we are obsessed with competing, and yet the perfect strategy is to find a market where there is no competition by creating a product that is unique and no one can copy.

MEGA TRENDS

In pursuing this strategy you also need to be aware of mega trends that are creating new demand. Technology, the environment, and health are examples of these influences. At any industry or business level there will be specific influences.

To understand mega trends look back in history and ask yourself what the eras of the last two thousand years were remembered for. "The glory that was Rome" was clearly about architecture and literature. When civilization emerged from the dark ages in the 1500s we think again of architecture and literature but fashion and clothing also become significant. In the 20th century we see the effects of the industrial revolution and mechanization. Planes, trains, and automobiles, which were really about communication. Today we are seeing the information age and technology being used to store and transfer information, or what we call IT.

However, other things are happening as a consequence of these advances in civilization, and a lot of them are negative. The chemical industry has a lot to answer for as a result of its behavior in the 20th century. I look back with disgust at the total lack of values, in fact destruction of values, in my education at the Chemical Engineering School in the University of Birmingham. Today's environmental movement to "save the planet" is addressing one hundred years of major abuse by the chemical industry.

A second major trend is related to this: the "social responsibility" movement. It's a consequence of improved communication. Globalization has led to people in the world's poorest countries being well aware of the lives of the "rich and famous." Equally, people in the developed world are becoming more conscious of exploitation in the third world. This fuels the equalization of wealth and affects issues like global terrorism.

Futurism has been around for a while and I do not intend to address it in this book. But one should be aware of it and also aware of the impact of future trends on your business. Books like *Megatrends* by Nesbitt[1] and work such as that conducted by the International Organization for Standardization (ISO) will point you toward high-level trends. Read *The Limits to Growth* by the Club of Rome,[2] another book that has shown an eerie truth in its predictions of the future.

YOUR OWN MARKET OPPORTUNITY

The question you have to ask is what is the next level of "trending" in your own market, and more importantly for yourself, where are people in your market having difficulty in "getting something done." It is easier to create a new product or service for the market you already serve than to create a new customer base. Look at your own market space first. You need to learn what changes are affecting your customer, both obviously and subtly, and the emotional effect those changes are having.

The accomplished innovator then looks for what is often called *white space* or *blue ocean,* which means "go to an uncontested market where there is no competition." Don't fight it out in the sea of blood called the *red ocean.*[3] You create a new demand and you provide a better product or service, also at lower cost.

A good example of a company that found uncontested space in an existing market is Home Depot. The post-2000 housing boom is an example of a *mid-level* business trend. Home Depot did not achieve growth in this boom just by taking business away from other hardware stores; it created a new market. The housing boom back in the 1980s led to a spin-off boom in home furnishings. People back then wanted to decorate, and the market was primarily female. The post-2000 housing boom was different. Technology had in the intervening years created new tools for the tradesman. The husband wanted to play with these new tools and people wanted to differentiate their house through more than just carpets and curtains.

Understanding the buyer's mind was critical. If guys played with these new tools they needed to understand the tools' capabilities. A power drill didn't just drill holes any more. However, be careful: this was the secondary market.

The primary market was improving the home. The traditional hardware store had focused on the secondary market. Home Depot chose its name because it was about the home and not about hardware. Their potential customers had two choices: hire contractors to do the work or buy tools and materials and do the work themselves. People choose hardware stores because they want to save money, but they often have problems doing the job themselves because of lack of skill or knowledge.

Home Depot created in-store clinics that taught customers skills like electrical wiring, carpentry, and plumbing. They picked easy-to-find locations and gave "friendly neighbor" service. Although their stores were warehouse-based they eliminated the misery of the self-service warehouse and showed people where to find things. They captured the enormous pent-up demand for home improvement. They made home improvement much easier to do.

In a totally different industry, in the 1980s the Polo brand recognized that the fashion market had split into the couture names such as Armani, Fendi, and Versace, and the commodity brands such as Gap and Hilfiger. The top end had gone glitzy and the bottom end had gone drab but pricey.

Polo has built its brand in the white space between the two market groups. It lacks creativity in design and charges high prices but it has a designer name, elegant stores, and most importantly, exquisite fabrics. Garments lacking fashion but made with quality fabric became its strength, and at a lower price than, say, Burberry.

Polo's timing was perfect to meet the business casual market of the '90s.

If you have a polo player embroidered on your golf shirt you can wear it to the office and everyone knows you must be a smart person.

Another good example of finding "market space" is in the accounting software market. Quicken could have focused on accounting firms for growth, but there are a lot less of these than there are small businesses. They saw that basic accounting packages were expensive, around $300, and full of accounting terminology. There was a growing market for small business and home business. Quicken retailed at about $90, created high value, and expanded the market by a factor of over a hundred. They made accounting easy for the small business.

THE FIVE FORCES

In defining where we will direct our strategy, Michael Porter's *five forces* model is still regarded as the definitive model for business strategy.[4] He sees five competitive forces that shape every industry (see Figure 5.1):

1. Bargaining power of suppliers
2. Bargaining power of customers
3. Threat of new entrants
4. Threat of substitutes
5. Competitive rivalry between existing players

Porter argues that the key to sustainable competitive advantage is either cost advantage or product differentiation, both of which relate back to the five forces model. He included a third factor, which is to be a niche operator, but even this will be either on a product cost or differentiation basis. His conclusion is that strategy is about choosing one route over another.

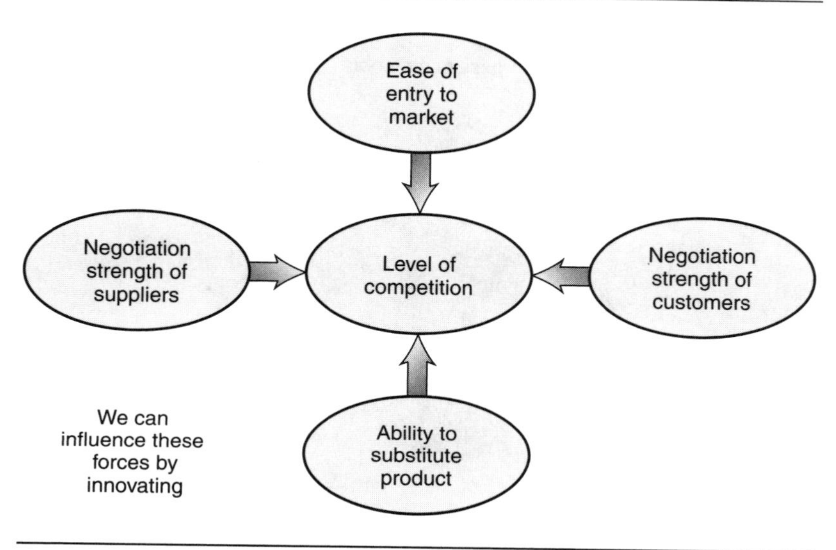

Figure 5.1 Competitive forces affecting survival of a business.

Too often in their obsession with being competitive, businesses slug it out on cost or on quality.

With globalization, competing on a cost basis is becoming less and less attractive. Porter, in his book *Creating and Sustaining Superior Performance by Competitive Advantage*, lists the following factors as affecting competitiveness.[5] I give you a typical example in each case:

1. *Entry barriers.* High capital investment and time to build, such as entry into heavy chemicals.
2. *Supplier power.* The oil industry suffers from raw material being located only in certain areas of the planet.
3. *Rivalry factors.* The retail industry is easy to enter and so is incredibly competitive.
4. *Substitution threat.* This is often subtle and the telecom industry, which for decades had a monopoly, now has to deal with this in a major way.
5. *Customer power.* People can walk into and out of a restaurant at will.

At the same time, you need to remember to "Do what you do well." In his book *Good to Great*, Jim Collins talks about the *hedgehog principle* and

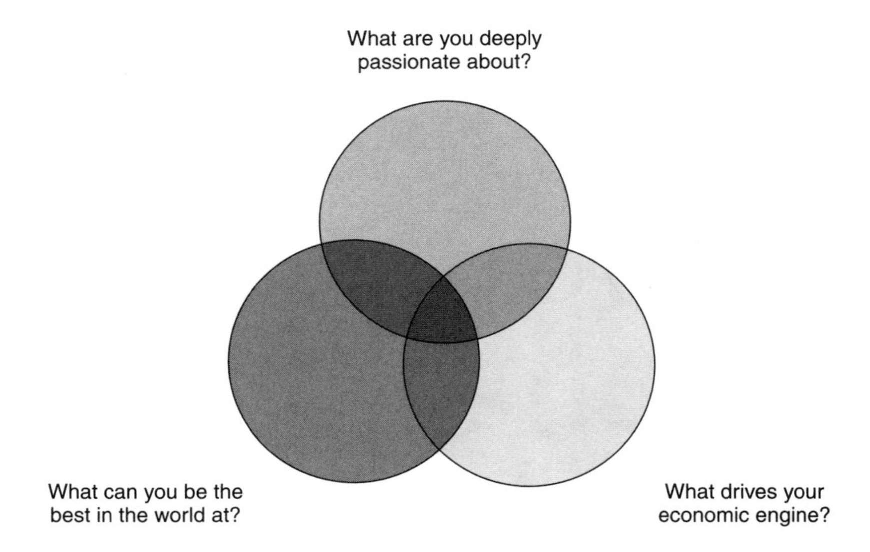

Figure 5.2 The hedgehog principle.

how successful companies pursue what they are passionate about and also do what they can be "best in the world" at doing (see Figure 5.2).[6] He also stressed that, at some point, you have to face the brutal truth that today's product is fast becoming obsolete and that you need to "innovate or die!" This means use your best abilities but point your abilities elsewhere.

START WITH YOUR OWN CUSTOMERS

Too often people will work on something they find exciting but that has no market relevance. More commonly they follow convention. Convention says compete in an existing market, beat the competition, serve existing demand, and provide either a better product or lower cost.

To initiate innovation strategy, you begin with your customer but talk about the customer's ideas . . . not yours. What are they having trouble getting done? Work with your worst customers and include sales as well as R&D in the discussion. You need to know what would make their life easier.

When the Swiss watch industry died, Swatch moved from jewelry to fun fashion. They still made watches. Yellow Tail Shiraz moved from the overpopulated dining market to the party market. They still made wine.

Business "hates"	Personal "hates"
1. E-mail crashes	1. Waiting in line at a store
2. Late starts to meetings	2. Slow restaurant service
3. Airport security	3. Waiting at the doctor's
4. Cancelled flights	4. Being stuck in traffic
5. Aggressive people	5. House cleaning
6. Shock hotel bills	6. A bad night on television
7. Declined Visa card	7. Noisy people in a book store
8. Boring conferences	8. Clearing snow
9. Technical support by phone	9. Power outages
10. Filing papers	10. Doing taxes

Figure 5.3 Author's business and personal "hate" lists.

Cirque du Soleil moved from the dying circus market to theatre. They still did the amazing acts. Do what you do well but pick your market carefully.

You need to know the top ten things your customers hate or that upset them. To get a good sense of what this means, make your own "hate" list. In fact, make two lists, personal and business. Here are my two lists of pet hates in no particular order to get you started (Figure 5.3).

Now use Figure 5.4 to make your own lists to describe your own reality. Off the top of your head list the things that have bugged you in recent weeks or months. One list is for your personal life and the other for your business life.

As you look at your lists you can probably see that someone provides some kind of solution to a number of your problems. Some of the solutions are expensive so you haven't adopted them, others are difficult to use so you can't be bothered. All of these problems are innovator's opportunities, and all of the solutions you have chosen not to adopt are innovator's opportunities.

You need to be asking the selfsame questions of your customers. What upsets them?

If you have problems with power outages you can buy a home generator for $5000, but you may prefer to buy candles and a deck of cards. That might be a good enough substitute. But if someone offered you a generator for $500 you might jump at it. You probably carry a backup credit card to avoid those "bad moments," and you probably ask for your hotel bill

Business "hates"	Personal "hates"
1.	1.
2.	2.
3.	3.
4.	4.
5.	5.
6.	6.
7.	7.
8.	8.
9.	9.
10.	10.

Figure 5.4 Reader's business and personal "hate" lists.

the night before you check out, in fact most hotels have "innovated" their service to do this anyway. We see increasing use of kiosks or ATM-type equipment to avoid queuing, but most of us don't like these because they are often badly designed.

It's the organizational problems that are the most frustrating: airport security, the appalling service from phone companies, or being stuck in traffic. These are instances where you feel you "can't fight the system."

You need to know where your customers feel they can't fight the system. It may point to a need to change your existing process. Unfortunately, well over 80 percent of so-called innovations fall into this category of minor process innovations. On the other hand, if you dilligently use your innovation process it may lead to a radical new offering.

Barely 15 percent, or one in six, of innovations can be described as "breakthrough" or "radical."

DECIDING WHERE NOT TO PLAY

Remember, in setting strategy you need also to decide "where not to play." This comes from staying with what you do well and using your core competencies. There may also be ethical issues such as gambling, environment, or health that influence where you "do not play." You can choose to play in whichever market works for you but again remember, it is much easier to develop a new product than develop a new market. Also remember that too much choice can be overwhelming, so determine which markets and products the company will not address. This gives a great sense of clarity as you clear away the "clutter."

In these early stages as you develop your innovation strategy you need to move away, but not too far, from the core business. You may want to look at the six to ten most successful innovations in your market's history and also the six to ten biggest failures. This can give you a good sense of direction. You then need to step out of your comfort zone.

This is OK for getting started as an innovator, but there are limits to today's customer's ideas and opportunities. There are more ideas outside the box. Open market innovation is your ultimate destination, which I will explain in Chapter 18.

Setting direction is done as an oversight to the first step of the innovation process, which will be about creating the opportunity.

BROWSER'S BRIEFING CHAPTER 5

- The perfect strategy is to find a market where there is no competition.
- In setting strategy we should note the high-level or mega trends that affect our existing and potential market.
- Understanding your market means understanding the mind of the buyer.
- The buyer is less interested in your product or service and more interested in what it will do to improve life for them.
- The "five forces" model of Michael Porter enables an organization to assess the impact of suppliers, customers, competitors, new entrants, and substitutes.
- Successful companies pursue what they are passionate about and what they can be "best in the world" at doing.
- Innovation strategy starts with finding opportunities with your existing customers.
- You find out where your customer is having difficulty; you do not go out and offer them your latest "cool idea."
- It is just as important to decide "where not to play" when selecting your target market.

Part II

The Process

6
Creating the Opportunity

Necessity is the mother of invention.

—Plato

It was December in Canada and we were all bracing for the winter. December is a month when you either catch up on your sins of omission for the year or wind down and take a break after working too hard all year. It was a bit of both when I met my good friend Bill Jappy for lunch. We always enjoy a bit of banter when we meet. Bill had Scottish roots and his family was from Aberdeen. My family was from Yorkshire and it was always a contest whether a Scotsman or a Yorkshire man was more tight-fisted. We liked to talk about football, that's the round ball variety, and his team, Aberdeen, were called the Dons and my team, Aston Villa, were called the Villains. It almost sounds like a meeting of mafiosi.

We met at a local restaurant called Alice Fazooli's, which contrary to the sound of the name was actually quite smart. We were shown to a table and sat down and ordered drinks. After the chit-chat we soon got into talking business. It had been a tough year. We were both involved in ISO 9000, Bill with a registrar and me as a consultant. My own past clients were moving into areas like Six Sigma or into my own area of specialty, which is of course innovation.

Bill, on the other hand, had limits on where he could move to. Registrars were in the business of auditing people to recognized standards and registering them to the standard against which they been audited. During the previous couple of years it had become a battleground, with the market saturated and registrars just stealing clients from one another—loyalty on a price basis.

There was not a lot to differentiate the service offerings of the various registrars, and many registrars were just looking for new standards to audit, such as environmental, food safety, or occupational safety standards. Bill was relatively new to his company so he had the benefit of a fresh perspective on the problem. He started to tell me about his solution to the problem and it was disarmingly simple.

Many of his company's clients were global operators and had the challenge of evaluating suppliers who might be on the other side of the planet. Bill's company had a global presence, with auditors located around the world. Supplier auditing is known to be a high cost in any company, and it is usually performed by people from within the company who perhaps do an audit once or twice a year and have limited experience and auditing skills.

Bill's company had auditors who were auditing every week and were highly skilled. With their global presence, travel costs were minimal and they eliminated the need for their clients to keep auditors on staff when they were not needed. Bill has created a whole new product for his company using the core competency of the company and moved it into an area of "white space" in the marketplace where they have no competition.

This is the stuff that innovation is made of. Thinking outside the box. To quote the fortune cookie saying, "The secret of a good opportunity is recognizing it!"

STAYING LOOSE

The first stage in the innovation process is creating the opportunity. People think innovation opportunities arise randomly. In truth, when you hear the word "random" it usually means, "I don't understand." This is a proactive activity and good ideas are not hard to find. The challenge is defining the problem in the marketplace. Let me explain.

You need to be looking for that perfect market where no competition exists. Years ago FedEx created a whole new courier market. The mighty Microsoft has so far been unable to unseat Intuit and its Quicken software in the small business market. Cirque du Soleil has created a unique market by innovating the tired concept of the circus. As I've said before, these are outstanding innovations. The innovator looks for what is often called "white space" or "blue ocean"—an uncontested market where there is no competition. You create a new demand and you provide a better product or service at lower cost.

I am a member of the innovation council of the Conference Board and we met Cirque in their Montreal base. Theirs is a remarkable story of innovation where Guy Laliberté and his friend were street performers in

Montreal witnessing the demise of the circus as we knew it. However, they saw a great new opportunity for more sophisticated entertainment for family and business. Cirque du Soleil was conceived. It's only when you see behind the scenes that you recognize how they have now created a market position that is very difficult to attack without a huge investment of not just money but time. It has taken Cirque 20 years to reach this position. They have also made an enormous investment in their artists, their training techniques, their design methods, and their relentless innovation.

It is interesting to compare Cirque with Yellow Tail. Yellow Tail is still the brand leader in its party niche, but their niche has been entered by a veritable menagerie of competition, including Long Neck from South Africa and Little Penguin, to name just two. It has been relatively easy for competition to enter this new market.

One brief vignette in this context: David Tallichet was a pioneer of theme restaurants and put his restaurants at harbours, hillsides, and airports. He used vintage airplanes to create atmosphere. He realized that people went out to eat in order to have fun together. He created "destinations." He brought fantasy to restaurants. Again, an easy market to enter, but one of the secrets of good marketing is to understand the market you are in. What does your customer really want? The customers of pharmaceutical companies do not want to buy drugs. They want to avoid being ill. In the third world the competitor of the pharmaceutical company is an organization like WaterCan, which is developing clean water supplies for the world's poor.

You must be watching trends and seeing the big picture, finding untouched markets that are creating new demand. But at the same time there will be specific influences at any industry or business level.

This "opportunity" work is done primarily by the creators, with some connectors involved so they can take the baton easily to the second stage. These people operate best in a loose or free environment. This allows the release of subconscious ideas that may have been stored for a long time. You should include a developer and a doer to keep your feet on the ground, and also so that they understand where these ideas came from when you get to the later stages of the process.

BE "OUT THERE"

You will find these opportunities by being "out there." Guy Laliberté's insight came when he worked on the street with his buddy. Exploration is a critical activity at the first stage in the process but so also is interaction with others. You don't have a new customer yet, only a disgruntled or disappointed customer in a dying market. Your mission here is finding the opportunity, and this may emerge only after months, even years, of searching. Note taking is the third critical activity at this opportunity stage. Look back at your notes on a periodic basis and you will be amazed at some of the "Aha!" moments that arise. I will talk more about this in Chapters 11 through 14.

Let me explain how to develop creativity and how to find innovation opportunities. I will start by telling you that you are more creative than you think! Let me explain.

BRAINSTORMING

During my "Introduction to Innovation" workshop I take people through an exercise that shows how you can be far more creative than you realize when you are in a group context. The exercise has a structure that owes a lot to the original ideas of brainstorming, which were developed in the 1970s and have been lost over time. There are three main inhibitors to good brainstorming:

1. A tendency to focus on one issue
2. Inhibition about providing revolutionary ideas for fear of ridicule
3. "Cruising"—a member of the group not contributing

The approach I am going to describe goes a long way to overcoming these problems.

THE SOCK EXERCISE

The group is asked to find alternative uses for a man's sock. The exercise takes about 10 minutes and it has five stages:

1. I first ask the group to write down three or four uses for a man's sock; each member does this on their own. Most people struggle to find four, the majority think of three. Examples might be: to polish furniture, or to use as a woman's sock, an oil rag, or a cash holder.
2. I then ask each person to turn to the person on their right and share their ideas and build their list. You now hear laughter starting. The list grows: as a bag, for a glove, to tie something, to light a fire, to mop a table, or to giftwrap something.

3. I then ask the people to turn to the neighbor on their left-hand side and again share their ideas. By now the buzz and laughter in the room are phenomenal. And the list grows again: as a muppet, a weight, a ball, a slingshot, a doll's hat, a pillow.

4. I finally ask each person to turn back to the person on their right and build a final list. They add a toy and a mask. From originally struggling to find three ideas, every pair of people now has a minimum of 10 ideas and some have as many as 12 to 18 distinct ideas, allowing for the fact that many of the original three or four ideas will have been repeated on each person's list.

5. We then capture the ideas of everyone in the room. We will have 20 to 30 ideas compared to the original list of about three per person.

This is collective knowledge at work!

I then ask each pair of people to select from step 4 one item from their list of 10 to 18 items that is the most unusual and one item that is the most commercial. Interestingly, these are rarely the same item. By the way, this is one of the common traps for innovators; we pursue ideas that are unusual and not ideas that are commercial.

You will notice that the technique I have described aims to find a new market for an existing product. Finding new market opportunities is partly what happens at this first stage of innovation, as I talked about in Chapter 5. However, finding opportunities to change your existing product is the other avenue you must pursue.

I will now tell you more of the rights and wrongs of brainstorming. Keith Sawyer describes brainstorming well in his book *Group Genius,*[1] and many others have revisited the brainstorming technique in recent years to restore it to its original effectiveness. Brainstorming is a technique you should use extensively in the first stages of innovation, so it is important that you understand it.

Note that the exercise I just took you through broke down idea generation into five distinct stages. Bear in mind the three main inhibitors to successful brainstorming that I described earlier:

1. *Excessive focus on one item.* Starting people separately avoids this.

2. *Inhibition.* Having people talk privately with two other people separately overcomes this.

3. *Cruising.* At every point each person must be totally engaged.

To be successful it is essential that team members listen to the ideas of others. The workshop exercise I described ensures that this happens. In large-group brainstorming, participants often tend to be busy trying to think of their own next idea or they "cruise" between contributions. By listening well we allow other people's ideas to trigger subconscious ideas and experiences from our own mind.

It is also important to recognize that creation takes time, whether it is a piece of music or a piece of software. We hear stories of bands creating a piece of music in an hour, but in truth that is probably from the convergence of a set of ideas they have worked on for many months. Each person in the band makes a contribution, and it is the same in good brainstorming.

You can also see that ideas need to be built "bottom up" to gain from the collective knowledge. What we call a breakthrough or epiphany will only occur after time. Groups will keep finding new problems for a long time before hitting on a solution.

WHAT WOULD MAKE YOUR CUSTOMER'S LIFE EASIER?

Customers are a rich source of knowledge when it comes to finding opportunities. However, you do not go to the customer and tell them you've come up with a really cool idea and wouldn't they love to have it! You don't ask what the new product is that they are looking for. The customer will think in today's context, not tomorrow's.

Instead, ask what would make their life easier. You need to be brainstorming with customers on questions like:

- What products or activities cause your biggest hassles?
- Which products or activities cause you to waste time?
- Which products or activities lead to problems being dropped on you?

These are not easy questions to answer because customers will have learned to live with many of these problems. From the answers you get, you define exactly what the problem is.

In Chapter 5 I showed you the opportunities some companies found. One more example I will share is Tumi luggage, who learned from their customers that ease of packing and being able to take suits straight from the closet was essential. While traveling, though, mobility was the key. Traveling people are now also carrying things like their laptop and techno toys. The growing female market meant that "any color as long as it's black" didn't work any more and also that luggage needed to be lighter in weight. This last requirement led Tumi to Boeing for lightweight frame tubes! But hold on, this is taking us to the second stage of the innovation process, the solutions.

THE PAIN STATEMENT

The output from this stage of the process is not some "fuzzy feeling"; it has to be a "pain statement." The problem must be given an initial definition. The definition will be refined in the next stage of the process but for now we need that pain statement that captures the unhappiness, real or subconscious, of the present customer who you intend to move to your new market. This is not easy because the customer will have gotten used to the pain. Henry Ford's customers were used to feeding their horse every day of the week when they only used the horse once a week. They had got used to paying vet bills. The horse manure was actually useful for growing crops.

The pain statement also needs to have some magnitude attached to it. I talk about measurement in Chapter 17 but for now think of metrics in terms of *problem frequency* and *problem impact.*

START WITH YOUR CORE CUSTOMERS

This builds relationships and provides the additional bonus that you keep out the competition. This develops your skills. You then move to your other customers, especially the customers you hate. They will "tell it like it is." Remember, finding pain is finding opportunities. This is where you find new needs. When you move to people who are not yet your customers you will find tomorrow's opportunities.

There are limits to the innovation that can be achieved by working with today's customer. Once your basic innovation process is working and you become more experienced, you then need to extend it to the *open market* concept where finding opportunities will then be much easier. These creation nets recognize that there is more knowledge outside the box. This is not new. In the Renaissance the textile industry in Tuscany operated like this. However, you must first learn how to learn, and you do this with your immediate customers! Open market innovation will also connect you to more potential solutions. I will talk about connecting in the next chapter.

Before you go to the next stage in the process, score your organization on how well it performs on this stage of the innovation process (see Figure 6.1).

Creating is finding the opportunity		Strongly agree	Agree	Disagree	Strongly disagree
1	Our people frequently come up with good ideas on their own				
2	We find out what problems our customers experience				
3	We interact with outside people to find new opportunities				
4	We explore outside the organization for market opportunities				
5	The work environment makes it easy to put forward new ideas				
		×4	×3	×2	×1

Figure 6.1 Innovation process assessment.

BROWSER'S BRIEFING CHAPTER 6

- Using the core competencies, and that includes knowledge, of your organization is fundamental to successful innovation.
- Using core competencies does not mean you have to continue to offer the same product or service.
- It is easier to change your product than to change your customer base.
- Finding opportunities is the job of the connectors, who operate in loose mode.
- Include connectors in this work so they can take the baton to the next stage.
- Brainstorming is one of the prime tools for finding opportunities.
- Brainstorming is not a loose activity; when done correctly it follows a well-defined process.

Continued

Continued

- We should pursue ideas because they are commercial and not because they are unusual.
- Creation takes time, and breakthroughs usually occur after a lot of hard work.
- Ask customers questions like "What are your hassles?" and "Where do you waste time?"
- Start with core customers, broaden to all customers, then move to "not yet" customers.
- Open market innovation moves us to "not yet" customers.

7

Connecting to the Solution

If I had an hour to save the world I would spend 55 minutes defining the problem and five minutes finding the solution.

—Albert Einstein

In a past life I spent a large part of my career in the Courtauld's corporation, which is now part of Akzo-Nobel. Although not a high-profile name globally, Courtauld's employed close to 100,000 people worldwide. In my early years I was with them as a chemical engineer. In later years, because of my artistic ability, I was chief executive of their Christy home textile brand. This was a powerful brand and had an 80 percent recognition factor in the UK.

One of my tasks was to visit the annual trade exhibition in Frankfurt, which was called Heimtextil. I would spend three days walking the exhibition halls viewing our competitor's products and over that time I would also gradually rediscover my ability to speak German. In the year I am thinking of I was specifically looking at duvet covers.

After three days of inspection and analysis I was finally leaving the snow, slush, and rain of Frankfurt and I was on my way home. I was walking through Frankfurt airport and my mind was still heavily focused on the products I had been viewing but it was starting to relax. I was starting to think about home. My daughters were still quite young and they would be getting ready for bed. Their beds would be covered in books as they read and settled down to sleep. If you have young kids you will recognize the picture.

A third factor then entered this melange. I found myself following a student wearing those big baggy dungarees. You know the type, made of denim and with lots of loose pockets. Suddenly I had one of those "Aha!" moments that I know you have had at some time in your life.

I connected dungarees, an exhibition full of duvet covers, and my daughters at home in bed reading books. I suddenly thought, "Wouldn't it be a great idea to make a kids' duvet out of denim with pockets on the side for them to put books in!"

Now I am sure as you read this you are thinking, "So what?" However, you have had this kind of "Aha!" moment yourself. It's exciting, and you want to rush off and share your excitement with someone. Sometimes you find solutions like this and you didn't even know you were looking for them! Well, I had to wait till Monday morning when I was back in the office to share my own excitement.

So far this is a classic story of how innovation should and does occur. On Monday that changes. I walk in the door. I am the chief executive and everyone courteously asks how my week went. Of course I say I've had this wonderful new idea and of course everyone says, "Yes, what a great idea." Again you have probably been in this situation.

This is not the way to move forward with innovation. All too often it is the boss's idea that moves forward. In truth, the collective knowledge of your people is far more powerful. Linus Pauling, who won two Nobel prizes, said "The best way to get a good idea is to get lots of ideas." He was a passionate believer in collective knowledge.

We actually went on to create and patent the duvet cover for kids that I had conceived. I am still proud of having that patent to my name. Nevertheless I am sure there were other better ideas in the minds of our people. The developers worked on the idea and improved it from my original concept, which is good. The fabric was made lighter and we used a technique called cross-dyeing to make it commercially viable. What I am describing here is the development stage of the product, which I will take you through in Chapter 9 and which I think we did very well. As a final point I would add that the execution stage, which I will discuss in Chapter 10, we did *not* do well.

DEFINING THE PROBLEM

In the previous chapter I talked about the pain statement, the frequency of occurrence, and the impact that are all part of the problem definition. Before we rush into burning energy on solutions we must identify the size and nature of the opportunity.

Home Depot saw an opportunity with the post-2000 housing boom, Yellow Tail with the party market, and Guy Laliberté in the entertainment market. However, what is the size of the opportunity? How many people are potential customers and how likely are they to adopt a new solution?

Describing an opportunity when the customer does not yet exist isn't easy. The even bigger challenge is not rushing to potential solutions, which we are all prone to do. Defining a problem in data terms avoids this risk. Attaching a financial figure really gains attention.

THE MYTH OF EPIPHANY

When we have insights people think the solutions come from some kind of epiphany. Epiphanies are in fact the last piece of the jigsaw puzzle and they arise from previously working on a problem. Archimedes's famous "Eureka!" came after working for a long time on how to establish the density of the gold in the King's crown. You can see why the loose mode is essential at this stage as well. Radical solutions come in all shapes and sizes.

If you look at history, most innovations build on previous experience, and most innovations also come as a result of stumbling through bad decisions.

If you want breakthrough innovation and want to avoid this stumbling, you should determine what aspects of your present product or service block the solution. This is where you sacrifice what have been regarded in the past as sacred attributes of your offering. The companies I mentioned in the "opportunity" stage all had the courage to do this. They all removed sacred attributes of their product. IKEA took out sales staff, Southwest removed first class, and Cirque du Soleil eliminated animals. However, they added something new after looking around and finding an unsatisfied customer need. IKEA added daycare and a café, Southwest added leather seating and entertainment, Yellow Tail Shiraz made their product fun, and Cirque du Soleil developed sophistication.

Going to the doctor is for most an inconvenience and out of their way. However, everyone goes to the supermarket. Target Stores started the MinuteClinic to deal with ailments such as pinkeye and strep throat. Who said you have to go to the doctor's office? People hated late or lost mail so FedEx replaced the train with the plane and developed a tracking system. Who said you can't track mail? These solutions don't seem radical today but at the time somebody surely said, "You can't do that!" Remember, you are "sacrificing the sacred."

This willingness to change requires the same loose or free environment of the creator stage.

This work on conceptual solutions is done primarily by the connectors with some creators staying involved. You should include some developers so they can take the baton easily to the next stage. Again include a doer to keep some sanity and also so that they understand how the idea developed in the first place. You can now start to see how the composition of the innovation team shifts as we move through each stage.

At this stage you must find and try alternative solutions. One is not enough. Remember, the best way to get a good idea is to get lots of ideas. You will stay with the concept and evaluate the concept. Call it prototyping if you like, but at a very conceptual level. You are not yet developing a working product or service. You will look at solutions that involve "moving context." Remember Henry Ford seeing that meat processing factory and getting the idea for automobile production.

There are a number of tools for doing this, one of which is TRIZ.

TRIZ

TRIZ was conceived in Russia in 1946 by Genrich Altshuller. His ideas were so revolutionary that he was imprisoned by Stalin, and the KGB used sleep deprivation as a way to break down his resistance. It is a testament to his inventiveness that he tricked his prison guards by cutting out two pieces of paper in the shape of his eyes, drew in his pupils, and placed them over his eyelids before going to sleep, so giving the impression of being awake.

After the death of Stalin he was released and went on to be responsible for many hundreds of patents. TRIZ has had a resurgence of interest in recent years as people have focused on the challenges of innovation.

TRIZ is pronounced *treez* and is a very "left brain engineering" method of solutioning. It can be said to be analytical and not creative. It has also been good at revealing trends such as the move from the mainframe to PC to laptop and from the power company to fuel cell. In that regard I see it as drawing on the laws of thermodynamics and specifically the theory of entropy, which addresses the tendency toward increasing disorder in a system.

There is a lot of highly technical material out there that is attractive to people who are uncomfortable with the creative aspect of innovation. However, unless you are an avid SPC fan, be careful. Rather like trying to use SPC outside the context of a developed quality management system, TRIZ will fail unless you use it in the context of an innovation management system, which was described earlier in this book in Chapter 4.

One of the better treatises on the subject is "Finding Your Innovation Sweet Spot" by Goldenberg and Horowitz.[1] They talk of many of the issues I have addressed already and how customers usually request only minor changes and how we try to overcome this with brainstorming and then eliminating radical ideas. They talk of how it is important to find ideas outside the business market and introduce what is a subset of TRIZ called *systematic inventive thinking*. This is thinking inside the box, and is based on *product elements* and *five generic innovation patterns*. These five patterns are:

- *Subtraction* where we remove components from the product as in the Slimline DVD player
- *Multiplication* in which we alter an existing component as in the move from the Gillette Sensor to the Gillette Mach 3 razor
- *Division* is the reconfiguring of existing parts as with the move from the "big box" to modular stereo components
- *Task unification* where we assign an additional task to an existing component such as with the combining of the car window defroster and radio aeriel
- *Attribute dependency change* is a little more complex and addresses the relationship between the product and its environment; an example of this is the photochromic lens that darkens in sunlight

This is for the left-brain analytical thinkers. The right-brain thinkers will use creative techniques like *I Ching*. I have used *I Ching* and I find it exciting; it definitely unlocks the subconscious mind.

I CHING

The *I Ching*, or *Book of Changes,* was written for use by ancient soothsayers for divining the outcome of their patrons' plans.[2] It was composed almost 3000 years ago and it was not until centuries later that the text was interpreted. It is pragmatic, and in looking at potential innovation directions it points you in a preferred direction. However, there is a need to understand the philosophy of *I Ching* before using it.

The basis of the book is a set of 64 six-line diagrams known as "hexagrams." Each of these symbolizes a different situation, and each of a hexagram's six lines symbolizes a different stage or aspect of that situation.

To make a divination, the diviner randomly sorts 50 yarrow stalks (I use kebab sticks) to pick a hexagram and put an emphasis on one or more of its lines.

I Ching is fun and it certainly frees the mind.

SACRED ATTRIBUTES

As a connector we are presented with a need and tasked with connecting to a solution. There are two key subtasks. First, determine what is blocking the solution, and second, looking at solutions in other contexts. Let's look at *solution blockers.*

I presented examples of companies who did this earlier in the chapter. They removed sacred attributes of existing offerings. A sacred attribute may have been needed 10 or 20 years ago when the offering was conceived or it may have just slipped in unnoticed.

Airlines are a good example of where social change and cost reduction have contrived to make the experience of flying today a total misery. It used to be that everyone was served a fine meal with smart cutlery, but with the onset of the TV dinner a cost-cutter saw the opportunity to save money. This worked for a while but the airlines' customers rejected TV dinners a decade ago and yet the airlines keep providing them. Serving a meal was a sacred attribute from the days when air flights were largely intercontinental.

The traveling misery was compounded by aircraft becoming larger (to achieve the consequent cost saving for the airline) but with the interior of the aircraft incapable of handling the logistics of the increased passenger load. The airlines further infuriated 80 percent of their passengers by cramping them into seating that was an ergonomic disgrace while giving 20 percent of their customers more room than was necessary. Airlines like British Airways even offer different attitudes and behaviors to the privileged 20 percent of their customers. This is based on the assumption that the 20 percent had influence and the 80 percent would not be heard.

It is small wonder that someone saw the opportunity. British Airways was attacked by Richard Branson with his Virgin Atlantic airline, and the many North American airlines were attacked by companies like Southwest and WestJet. The "new market" carriers stopped providing bad food and cramped, uncomfortable conditions but did provide entertainment and friendly service.

CONNECTING

This means connecting to social shifts that impact your product, and also connecting to other business environments where these social trends may have already been picked up. Networking, which I describe in Chapter 14, is one important way of doing this, but the fundamental point is that breakthroughs occur at the intersection of bodies of knowledge—the "spark of ingenuity." This is about understanding the customer experience. My friend Teri Yanovitch, with whom I worked at Crosby, has written extensively about the customer experience in her book *Unleashing Excellence.*[3]

Some of the social shifts that can influence products, as I mentioned previously, are seen in Home Depot, FedEx, Yellow Tail, and the short-haul airlines. In retail, Home Depot saw their business as similar to IKEA but different from IKEA in the technical component and need for a high number of stock-keeping units (SKUs). The social move toward one-stop shopping meant they needed to provide "friendly guidance." FedEx saw the globalization opportunity that the Post Office was not addressing and connected it to warehousing and bar codes. FedEx operates a worldwide warehouse. Delivery is what the customer sees—warehousing, whether on a plane, on the ground, or in a truck, is what FedEx does.

I suspect that Yellow Tail Shiraz connected to the social move toward drinking more fruit juices; I wonder when the airlines will see the social move toward drinking coffee in a pleasant environment. When will that short-haul air flight become like sitting in Starbucks or Second Cup?

The connecting I am describing recognizes the universality of knowledge and that someone somewhere has probably already thought of the solution you are looking for, but in a different context. That is the power behind networking for collective knowledge.

WALK BEFORE YOU RUN

Sow the seeds of your innovation process at a level where you know you can succeed. I don't want you to stay locked in process innovation but it is a good place to start. Let me tell you about one of my clients, Jim Laforet, who talked about how he started the innovation process in his own company, Spectra Energy, at a recent "Breakfast of Innovators" that I run on a periodic basis. Jim manages the customer service operation at Spectra, which is a gas distribution company. As Jim said, it is hard to be innovative with a molecule of methane. Add to that the fact that regulators do not reward innovation. However, Jim had "innovation" on Spectra's performance assessment and in the mission statement, so he brought his 24 supervisors to a two-day workshop that I put on for them. The workshop was similar to the one I describe in Chapter 13, but with extra time to actually start work on opportunities.

The group produced 45 innovation opportunities of which 22 were implemented. The changes varied from restructuring of meter routes to simple things like paper shredding, but he got his people thinking in "innovation mode." The fundamental in all of this was that customer satisfaction impacts revenue in a call center. Happy customers pay on time and complain less, whereas unhappy customers do the opposite.

What Jim's team learned in the workshop and in subsequent implementation was that people needed to know their innovation roles. Innovation was a state of mind and you have to keep it fresh.

They also saw that they could apply the innovative state of mind to their processes, and their "early win" was the elimination of 2000 hours wasted handling high-cost bills in a process they changed through rapid innovation.

You must constantly be testing new ideas, some of which may even be in conflict. The ideas are not random; they will be selected based on trends and the core competencies of the business. You are testing future opportunities and constantly trying out different options in the market. You find what works, stop the things that don't work, and resource the things that do. Don't wait for the competition to make your products obsolete, do it yourself. This is done in conjunction with your business strategy activity. The knowledge gained from this testing is a prime input to strategic planning, which I discuss in the next chapter.

You can see that if you approach this second stage of the innovation process in the right way you will generate lots of ideas. The challenge is then, "Which ideas do we run with?" This takes us to the tipping point in the innovation process where we select our solutions.

Before you go on to the next stage in the process, score your organization on how well it performs on this stage of the innovation process (see Figure 7.1).

Connecting is finding the solution		Strongly agree	Agree	Disagree	Strongly disagree
1	We work with clients to find solutions				
2	We find markets where there is no competition				
3	We use a defined solutioning process for our opportunities				
4	We find solutions by going outside the organization				
5	We find the best solution				
		×4	×3	×2	×1

Figure 7.1 Innovation process assessment.

BROWSER'S BRIEFING CHAPTER 7

- We all have the ability to be connectors.
- Using collective knowledge is the most powerful way of finding alternative solutions.
- Most innovations are built on previous experience.
- Be prepared to sacrifice sacred aspects of your current offering.
- Somebody will be sure to say, "You can't do that."
- An epiphany is really the last piece of the jigsaw puzzle.
- Connectors find conceptual solutions and include developers who can take a solution to the next stage.
- TRIZ is a systematic approach to finding solutions and is preferred by left-brain thinkers.
- Right-brain thinkers prefer tools like the *I Ching*.

8

The Tipping Point

I shall be telling this with a sigh somewhere ages and ages hence: Two roads diverged in a wood, and I—
I took the one less traveled by, and that has made all the difference.

—Robert Frost

I am going to say that risk taking is different from gambling. Risk taking is where you assess the likelihood of an event occurring and you assess its impact if it does occur. If you want to innovate you have to take risk. However, risk taking does also have an intuitive component.

I am a senior advisor to Toronto ASQ, and each year the section treats its "helpers" to a night out for their contribution. We have been to some interesting places. A couple of times we have taken a boat cruise on Lake Ontario. We have been to the theatre, and the year I was chair I recall we had a night at one of the city's jazz clubs.

More recently a really good friend of mine, Lucy, was chair and she arranged a night at one of the local casinos. Well, to quote the title of a show, "I love Lucy" but I couldn't bring myself to be there and yet I desperately wanted to support her. The event made me realize what a strong aversion I have to gambling, and yet more recently I read a questionnaire that indicated that you can't be in business without being a gambler.

My thoughts then took me back to my younger years when I used to play a lot of Monopoly. They say Monopoly brings out your dark side; I recall the high stress I used to feel over what was "only a game." I used to win a lot because the game contains some marginal decisions, and yet from time to time no matter what I did I couldn't help but lose. Maybe I got stuck "in jail" while everyone was picking up properties or just kept landing on properties that someone else owned.

Innovation is where you take the marginal decisions and take risks like in a game of Monopoly.

While I don't gamble, I do take a lot of risks, but they are measured risks and also managed risks.

SELECTING SOLUTIONS

Malcolm Gladwell's book *The Tipping Point* is a great read and its title has introduced a major new term into the world of business. *The Tipping Point* points to how small occurrences have a major impact on subsequent events.[1]

You are taking major risks at this tipping point in the innovation process, and it is critical that the attached risk is both measured and then managed. Although I will repeat, *do not deny your intuition.*

Gladwell's *Tipping Point* is about how an idea, trend, or fashion crosses a threshold and spreads like wildfire. That's the point we are now at in the innovation process. The selection of which opportunities and solutions we will pursue is the "tipping point" in the innovation process.

There are four common failings in making these selections:

1. We do not make a selection, and then we pursue too many options and so under-resource our options.
2. We select without good data and make choices based on the wrong criteria. An idea may be unusual and gain attention but have no commercial merit (remember the sock exercise in the earlier chapter).
3. We select in a risk-averse mode and pick only short-term options where we are competing in a crowded marketplace.
4. We only assess risk internally and not externally (I will talk more about that in a moment).

Remember that statistics show that out of 3000 ideas only one will make it. Unfortunately we often kill the best ideas with overaggressive requirements for ROI. On the other hand, we often let a high-risk product through because we failed to assess the risk attached to a potential new product. We need to make the right decisions on where to invest our resources.

SELECTION AND STRATEGY

The strategic planning sessions of your organization are where you make the decision on what to select. The high-risk "accidents" from the first two stages of the process must be part of the input to the session. We also need enough data and information input to this session to be able to select based on risk and make decisions on which products to move forward. In the packaged goods industry 30,000 products per year are introduced and between 70 percent and 90 percent don't stay on the market more than 12 months. This is partly due to test marketing but also due to a lack of sound risk assessment.

Many CEOs complain that their strategic planning process does not produce new ideas and is often just a game of politics. The annual strategy review is often a series of adjustments to last year's numbers with very few new ideas. It does not prepare for the risks ahead or serve as the focus for the company's direction. Strategic planning must actually make strategy and encourage risk taking.

There must also be quarterly reviews of strategy in which the mix of the portfolio is checked and the expected ROI and risk are also checked. This reduces the feeling of the annual review being a "once a year only" activity. This is done by your core group of strategic planners, who are usually rising stars in the business.

SHORT-TERM THINKING

We need enough data and information input to the strategy session to do risk assessment and enable us to make forward-looking decisions.

Return on investment is good as a guideline but not to rank ideas for a decision. Just determine whether the revenue will be a hundred thousand or a hundred million.

The other thing to beware of at this decision-making point is short-term thinking. Most companies look for three years' ROI from any new product. Xerox, one of the great innovators, has found that it has had to wait an average 7.5 years for an acceptable ROI on its best innovations. In a different market, the cell phone was launched in 1985; widespread payback did not come until 1998.

De Smet, Loch, and Schaninger in *Anatomy of a Healthy Corporation* talk of the pressure with which we are all familiar.[2] The pressure for short-term performance and the tendency to react to what is "in your face" overriding the need to invest in long-term thinking. Their research shows something that intuitively we accept but in practice we all too often reject. The portfolio of innovations in which we invest must contain a mix of short-term certainties and long-term risks (see Figure 8.1).

The low-risk projects of two years or less are very much driven by immediate customer need. These rarely qualify as breakthrough

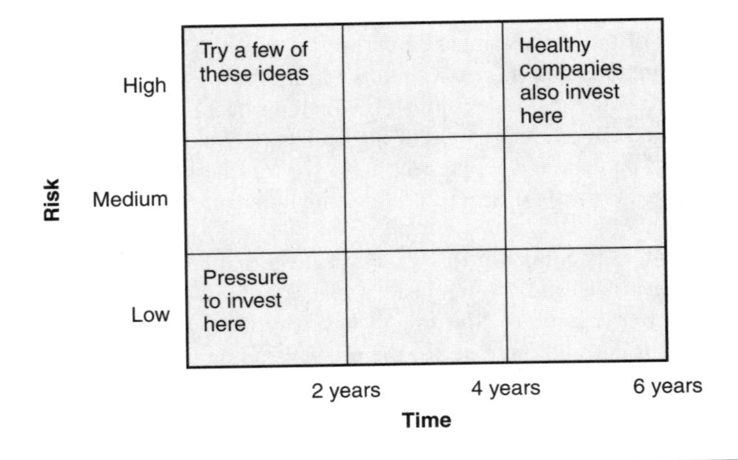

Figure 8.1 Project risk versus time scale.

innovations. However, the strategists must also invest in wild-card opportunities in the short-term, where the risk is high and learning will be fast. Financial institutions such as Capital One are becoming good at this. You must be prepared to fail.

Your portfolio must also include the longer-term five- to seven-year projects, both low- and high-risk. This is your "seed corn" for tomorrow's business. These don't all have to be high-risk. The long-term selection comes from viewing market trends. My good friends at Canadian Standards Association decided several years ago to invest in developing a guidance document on global warming. Some people looked at this as a make-work project. The product was developed and completed and then six months after completion, Al Gore's *An Inconvenient Truth* hit the media and CSA were in pole position at the racetrack. Some would say that was good luck. I would argue that this is looking at the significant trends in your industry and recognizing the core competencies of your own business. This is not just a job for the CEO; it is a cross-functional activity.

What are your own strategic influences? The economy, China, broadband, the environment? You need to be evaluating the long-term impact on your business.

BEWARE OF "MINOR" INNOVATIONS

Business in the '90s became obsessed with "minor" innovations and deluded itself that it was being innovative. This was all under the mantra of continuous improvement. Business has become obsessive with squeezing the last drop of juice out of the lemon of efficiency. Meanwhile the world has moved on and your market has changed. Research by George Day has shown that through this period the percentage of major innovation in portfolios dropped typically from 20 percent to a little over 10 percent.[3] We became risk averse. On the other hand, the same research shows that in a similar period, 60 percent of the profit from innovation came from the major innovations. Clearly we must aim for major innovations. This means moving beyond the "baby steps" I described earlier in this chapter and starting to be bold.

Major innovation means a product and/or a market that is new to the company. We just looked at risk versus time, and clearly risk increases the further out in time you are projecting simply because you predict future events with less and less accuracy the further out in time you go.

Risk also increases as you become less familiar with the potential product or market (see Figure 8.2).

The higher the risk, the higher the chance of failure. However, remember, Intel says if they are not getting ten failures to every success they are not taking enough risks.

In order to assess whether your product and market involve radical change you need to ask yourself the questions in the survey shown in Figure 8.3.

Your maximum score is 20 in each case. Low risk is less than seven points and high risk is more than 14. You will keep revisiting these

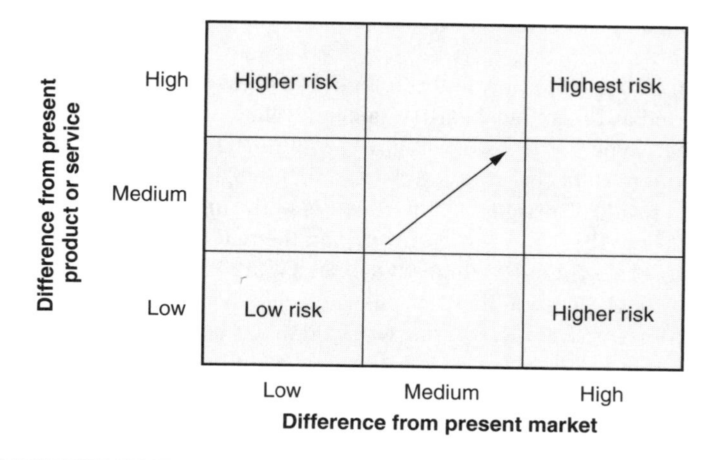

Figure 8.2 Project risk versus familiarity.

For the product	Strongly agree 1	Agree 2	Disagree 3	Strongly disagree 4
We have the knowledge/technology				
We have the development capability				
We have the competencies				
We have IP protection				
We are capable of production/delivery				

For the market	Strongly agree 1	Agree 2	Disagree 3	Strongly disagree 4
We have the business (supplier) partners				
We have the delivery mechanisms				
We have early adopters				
We know our competitors' positions				
We know the customer decision process				

Figure 8.3 Radical change assessment survey.

questions as you conduct your quarterly review. This will show you how risk is shifting.

You also need to look at the potential ROI attached to the risk you are taking, and as I mentioned earlier look at revenue in order of magnitude.

Clearly your perfect world is low-risk/high-ROI, and the nearer you are to the bottom right corner of the chart in Figure 8.4, the more attractive the project. I would also argue that in the world of the innovator, the bottom left (low-risk/low-ROI) is a waste of time and the real challenges come in the top right corner (high-risk/high-ROI). See Figure 8.4.

The tough question in all of this is when, whether, and how to kill an idea. A sterile risk analysis is one way, and this is better than a simple ROI. The question is more *when* to kill it than *whether* to kill it. The tools I have just described as part of the strategic selection process need to be fueled with data that were developed during both the opportunity and solution stages. The tools themselves should be used as assessment guidelines during the earlier solutioning stage.

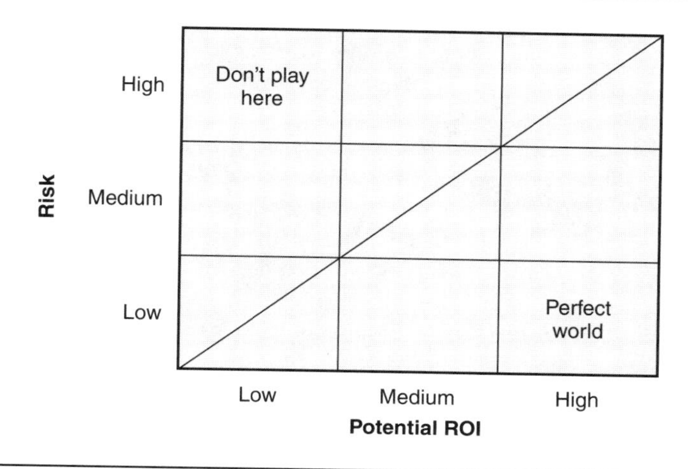

Figure 8.4 Project risk versus ROI.

INTERNAL AND EXTERNAL RISK

One other perspective on risk needs thought and that is internal versus external risk. We usually assess internal risk well because we are close to it, but external risk has far more impact because we are less able to mitigate it. One thing that Edison and Bell both did well was take care of the infrastructure that their new ideas, the light bulb and the telephone, both needed. Whereas Michelin with its "run flat" tire and Sony with the Betamax hit major problems with great products because the infrastructure was not in place to deploy the product. Risk needs to be evaluated in the present selection stage and then addressed in the development stage, which is discussed in the next chapter.

The first external risk is with suppliers and subcontractors. Do you know who they will be? How reliable are they? Do you have backup? The chances are that if you have a new product or service then you have at least one new supplier or subcontractor, and they may not be conventional. It may be software or knowledge that they are supplying. All your other suppliers can keep their promises but if this one fails, you fail. Assess your supplier. If you have a QMS you know how to do an audit. Evaluate their management system and their design and development processes. If you don't have the audit capability, get an ISO 9000 registrar to do it for you. If there are weaknesses, get backup.

If you have four key suppliers with three of them at 90 percent probability of delivering and the fourth at 40 percent probability, then the joint probability of your supply base coming together for you is only $0.9 \times 0.9 \times 0.9 \times 0.4 = 29\%$! You certainly wouldn't move forward with an internal risk at that level without serious mitigation.

At the other end of the business you must assess how many hurdles you must jump before you reach the ultimate user. The classic modern story here is Michelin with its "run flat" tire. They began development in '92 and introduced it in '97. The original equipment manufacturers (OEMs) had not been close to the development and so then needed another three years to include it in a new vehicle from the design start. The OEMs were cautious and so introduced it in a limited number of vehicles. On top of this, garages needed special equipment and staff training, so by 2005 the product had only limited use on a restricted number of vehicles.

Who is your distributor? What is the decision time cycle? These numbers impact the ultimate time to market. If you are entering a new market, these cycle times will need even more analysis than your supplier capability assessment. This data is going to be far more difficult to obtain, and of course as you start asking questions it gives a heads-up to your customer and also to your competition. Maybe you need a business partner with market knowledge in your intended new market.

The lesson? You may well find it better to pursue a new product or service with high internal risk that you can manage and mitigate and a low external risk that you don't have to manage.

Having made your decisions the game now changes.

SWITCHING FROM LOOSE TO TIGHT

Once decisions have been made, we switch from loose to tight. The tipping point is where you actually change your mode of operation. So far this has been a loose and flexible process, but now we need to make some firm decisions so we go tight. In the first two stages you will find lots of opportunities. Now you narrow your focus. A key task for leadership is monitoring this change in mode. We now need to move with speed.

Again, before you go on to the next stage in the process, score your organization on how well it performs on this stage of the innovation process (see Figure 8.5).

Strategic planning is the tipping point		Strongly agree	Agree	Disagree	Strongly disagree
1	We do not go for short-term ROI				
2	We assess external risk				
3	We assess internal risk				
4	We have sufficient data to make decisions				
5	We provide resources to support our decisions				
		×4	×3	×2	×1

Figure 8.5 Innovation process assessment.

BROWSER'S BRIEFING CHAPTER 8

- Risk-taking is fundamental to innovation.
- Risk can be measured, but there is an intuitive component.
- The tipping point in the innovation process is where we select which options to pursue.
- Organizations frequently try to pursue too many options.
- Choices are often made based on poor data.
- Organizations have become risk-averse and only pick short-term options.
- External risk is often assessed poorly.
- A new-product portfolio must include a number of long-term options.
- There is a danger of being drawn into "safe" and minor innovations.

Continued

Continued

- A major innovation is a product or market that is new to the company.
- Risk is reassessed as you move through the development stage.
- External risk is more difficult to manage than internal risk.
- Risk exists both with partners (supply chain) and customers (delivery chain).
- This is the point where the process switches from loose to tight.

9
Developing the Solution

Genius is one percent inspiration, ninety-nine percent perspiration.

—Thomas Edison

Above everything, don't fool yourself into thinking that because you have found an opportunity and also found a solution that there isn't someone else on the planet who has found the same opportunity and probably a similar solution. To quote the lawyers, "time is of the essence."

Alexander Graham Bell is probably the classic story of "the need for speed." He worked for many years with his colleague Thomas Watson developing a way of being able to send "sound over the wire" and being able to "talk with electricity." He was also well aware of patent requirements for intellectual property through his father-in-law to be, Gardiner Hubbard, who ran the Boston patent office. It is one of those legends of business history that he beat his archrival, Elisha Gray, to patent his device by only a matter of hours.

In most businesses, the market opportunities are actually fun to find. It is the development process that is the tough, grinding work. Let me quickly revisit the early stages of the innovation process in the context of the business I used to run.

In home textiles the market barometer was the clothing industry, and when I was in the clothing industry the barometer was always the couture collections that would push the envelope in color, garment structure, and fabric pattern. Many trends in fashion are cyclical and also relate to social mood and the economic cycle. At Christy we would source our own graphic ideas from artist colonies such as the one beside Lake Magiore in Italy. Let's say, for example, we saw an opportunity in paisley prints. We would look for alternative treatments by different artists. Remember, "The best way to get a good idea is to get lots of ideas."

Having gone through stage 1 (opportunity) and stage 2 (solutions), we would then select. There is an overlap between selection and development as there is with the other stages in the innovation process. At the selection stage there may be a need to test alternative solutions, and you may choose to carry forward more than one solution. You may pick two alternatives for development and very rarely pick three. This is where many people go wrong and take too many choices into the development stage.

My chief designer was Lorraine Robert from Montreal—a colorful lady, but as tough as nails. She had arrived in England via New York and France. She was a brilliant artist and I balanced her creative ability with a sense of market need. She also worked incredibly well with customers. Finally, I also brought the skill of color into the equation. They were exciting times. We would return from Magiore with our plunder and lay out the artwork for the upcoming development work.

Having selected the one or maybe two paisleys we wanted, we would run with the first step in the development process, which was to transpose them into a form that was 1) user-friendly and 2) could be manufactured easily.

You need to ask what these two things mean for you. User-friendly for us meant it would coordinate with color schemes being adopted by the consumer but at the same time give them a new and exciting experience. Ease of manufacture was the other critical consideration for the developer. Each color we added to the design increased cost, but one color is boring, two becomes interesting, and three starts to get exciting. A five-color print is extremely expensive and also fraught with the risk of complexity. The magic solution is of course one color that is exciting.

In your own world ask whether you are falling into the trap of too much complexity in your design and whether the service or product you are developing will be easy for your customer to use.

The irony of this development stage in the innovation process is that it is usually the most defined stage for most organizations and yet it is where most innovations get killed. One of two things happens: either the development stage is so slow that someone beats you to the market or there is a tendency to revert to status quo and kill the new idea.

The development stage receives its input from the connectors after the opportunities have been screened by the strategists. Companies pour monies into development labs instead of into the creating and connecting stages, which are actually more fundamental. Even companies that spend the most on development are starved of innovation and market growth. Development is an area where many companies are well resourced but are simply producing more of the same.

A key to success here is having a well-designed development process. Whether you are designing a chemical plant or creating a fashion item, the stages are similar. I am going to take you through the stages for a new graphic design in textiles. You will relate to it as you are a user of the product and it is a fairly simple process.

THE DEVELOPMENT STEPS

Whatever the service or product your organization excels at providing, you have certain distinct steps in designing that service or product. Those steps must be clearly specified along with the "entry and exit" or "input and output" criteria for each step in the process.

Let me take you through the steps of developing a new print design. While you are reading this book you are probably surrounded by prints, whether in the fabrics in an airport lounge, the magazines in a coffee shop, the furnishings in your own home, or the clothes you are wearing. Look at one of these prints and let me tell you how they were created:

1. The original concept artwork will have been created manually or by computer graphics. Even with computer graphics the artwork will normally need to be printed as a hard copy to show the color of the ink in true form.

2. The design will be extended to evaluate how it looks "in repeat." A graphic that looks good in a 12" square may not work when it repeats over an area six foot by six foot (36 square feet). The repeatability of the design is addressed.

3. Color normally comes next. The concept will be analyzed by the design team to select alternative colorways. The colorways will not be final, as new colorways will emerge during prototyping. Color balance is critical, and the ability of new colors to work with existing domestic colors is also vital.

4. The fabric strike-offs are next. How will the design behave on the final fabric to which it will be applied? Lighting effects like *metamerism* must be addressed; a particular pigment undergoes a color shift between daylight and artificial light.

5. Once satisfied with the strike-off, a short production run will be required. Depending on the printing technology, this may require investment in components for the printing equipment. This can be a significant scale-up cost. This enables identification of production problems and also enables sufficient fabric to be produced for a short production run of garments.

6. The short run of garments is tested on a cross-section of customers, who will have been involved at earlier stages but not in detail. They may not have the visualization skill to see strike-offs in the final garment.

Clearly, you want as few iterations of this cycle as possible and you want it to be as fast as possible. Recording design changes as you go through this process is something developers have learned from bitter experience. What developers do not do well is conduct a detailed design review to ascertain whether requirements have been met at the end of each development stage, and also concisely record the key issues at the end of each development stage.

I have deliberately taken you through a simple example so that I can explain concept and approach. You may be saying, "What has this got to do with designing a chemical plant or a new service delivery method?"

Look at the steps in the process and apply them in your own environment. It is essentially 1) concept, 2) scale-up, 3) test, on a repeating basis. Recording decisions and actions at each stage is vital to avoid reinventing the wheel in your development stage.

THE CHANGE FROM LOOSE TO TIGHT

You can see that this stage in the innovation process requires discipline. It is often a challenge to shift from the fun and freedom of the early creative stages to the data recording and analysis of the development work. For me it was a shift from the freewheeling and wining and dining of Lake Magiore into the grind of dimensioning, cost analysis, and meeting deadlines. This is a vital shift you have to make.

This stage should be fast and not secretive, and should involve the ultimate user. Remember, this is where you move from loose to tight. Good project monitoring and control is vital.

The job for the developer is to make the solution work and make it user-friendly.

This development stage also requires the building of partnerships with your customers and distributors to enable the later operation of the delivery chain. This partner management must use data from the risk assessment conducted at the tipping point in order to manage relationships where the risk is highest.

This work is done primarily by the developers, with some connectors and creators staying involved to ensure integrity. You need to be including more doers by now so that they can pick up the baton easily in the last stage. However, you will see a tension building in the "I" team (innovation team). You will recall how developers think "creators are not focused" and creators think "developers don't see the big picture." Keeping a team mix is important and leads to better results.

BEHAVIOR CHANGE

One of the biggest obstacles to acceptance of new ideas is behavior change on the part of the user. Research has shown that sellers overestimate the perceived benefit of a new product by a factor of three while buyers overestimate the perceived benefit of an existing product by three times. The barrier to the new product ($3 \times 3 = 9$) means that to overcome this barrier we need a $10 \times$ increase in perceived value—an exponential increase.

Most new ideas struggle to find daylight because potential users resist anything new. The challenge for the developers is to make the new product or service so user-friendly that resistance is futile. The job for the developer is to make the solution work. Let me explain some of the challenges.

The psychology of behavior change is worth understanding. The classic approach in business to overcoming resistance to change is to create an attractive vision of the future. The word "resistance" implies that we have to "push." We will talk about creating the vision or *value proposition* in the next chapter, but first we need to put ourselves in the mind of the potential customer.

The book *Managing Transitions* by William Bridges brings out the issue very well.[1] In creating so called "resistance," people are actually saying there is something about the status quo that makes them very happy. We may have found what we regard as the coolest solution on the planet but people will not go through unnecessary "pain" to adopt our solution. In June 2001 *PC World* magazine reported that over 60 percent of MS Office users were still using Office 97 and even Office 95, and Microsoft had taken the drastic step of no longer providing free support for these versions in order to pressure people to upgrade.

The developer's job is to address the pain of change and make the new product user-friendly. It is not to tinker and build in bells and whistles that the developer personally finds attractive.

RESISTANCE TO CHANGE

Daniel Kahneman won a Nobel Prize in economics in 2002 for his study of why people ignore what appear to be obvious benefits, whether in time or money. People actually look at a new innovation from the following points of view:

1. Perceived value, what they think it is worth.
2. From a reference point, comparing to something they know.
3. What are the gains and losses they will have.
4. The opinion that losses matter more than gains!

This last point that losses matter more than gains is most important of all.

The research by Kahneman showed that people buying a product would typically have the mind-set of "just hang on to what you've got."[2]

An example of this resistance to change is the Dvorak keyboard. It increases typing speed by 30 percent but who wants to relearn how to type? The ketchup bottle with its opening at the base seems like a great idea. The manufacturers even apply the label upside down to show us which way to stand the bottle. Wait for this . . . in the first five years of its life 70 percent of people still stood the ketchup bottle with the opening at the top and the label upside down! People do not change easily. See Figure 9.1.

If you want an example of an innovation success, look at Google. Who wanted a new search engine in the year 2000 with so many failures around? Google's algorithms are amazing, and hence its success. The user interface is so simple some people no longer record Web site addresses. No behavior change required!

We can all recall examples where our own resistance to change slowed down adoption of something.

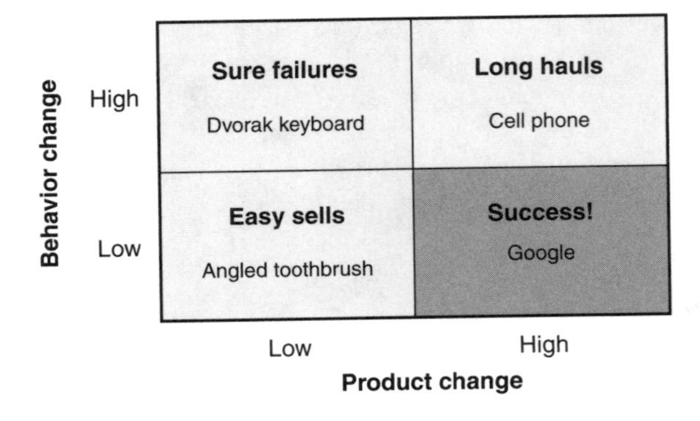

Figure 9.1 Behavior change and innovation.
Source: John Gourville, "Eager Sellers—Stony Buyers," *Harvard Business Review* (June 2006).

I recall my wife buying me one of the first digital cameras from Sony. It was so difficult to use I continued to use my old roll film camera for a couple of years. This actually saved Eastman Kodak, who hit the market with a camera that embraced the simplicity of the old box Brownie. It was so easy to use, and I recall my friend Trevor Smith, Global Quality Manager for Kodak, demonstrating to me the simplicity of their product.

Don't ever allow yourself to go to market with a product that is too complex to use. Simplicity and ease of adoption are your driving forces here.

Another way of looking at this is to aim for exponential improvement or benefit, a factor of ten. Think back to that research of Kahneman's.

If you are aiming for "white space" or "blue ocean" then you are creating a new market and you might think you have no resistance, but you need to think carefully. The iPod has opened the market most easily with people who have not yet bought a car or a home stereo, the snowboard with people who have not yet skied, and the mountain bike with people who do not own bicycles. But this is more for the next stage in the cycle when we address the value proposition of innovation.

Before you go on to the next stage in the process, score your organization on how well it performs on this stage of the innovation process (see Figure 9.2).

Developing is making the solution work		Strongly agree	Agree	Disagree	Strongly disagree
1	We fund novel projects				
2	We are willing to take risks with new ideas				
3	We are able to get people who can work on new projects				
4	We closely monitor project progress				
5	Our projects finish on time				
		×4	×3	×2	×1

Figure 9.2 Innovation process assessment.

BROWSER'S BRIEFING CHAPTER 9

- At the development stage, time is of the essence.
- The chance is that someone else has found a solution that will compete with your own.
- Development is overfunded and underdelivers in far too many organizations.
- A well-defined development process is fundamental to success.
- The input and output criteria at each step of development must be clearly specified.
- The essentials of development are 1) concept, 2) scale-up, and 3) test.
- Organizations often have difficulty moving from loose to tight mode.
- Partnerships with the supply chain and delivery chain are developed here.
- The primary challenge of the developer is to make the new product or service easy to use.
- Potential customers will significantly overvalue their existing product and undervalue the new offering.
- You will generally need an exponential improvement in time or cost savings for rapid acceptance.

10
Execution

Time is of the essence.

—An honest lawyer

Over a century ago people in Europe and North America used what we today call "huckaback" towels, which were a flat cloth with a honeycomb weave. The Christy family was a typical wealthy Victorian family with each of the sons having a destined profession and any daughters being pointed toward other wealthy families for an "appropriate match."

Henry Christy was one of the sons of William Christy. He was sent on what was a statutory journey of the time. It was called the "grand tour." It was a journey across Europe and its aim was to broaden the mind. Henry attained the outer limit of his tour when he reached Turkey and through his connections was able to meet the sultan.

Henry was given a tour of the royal palace and that included the "ladies quarters" or as we call it today the *harem*. He saw the ladies hand-weaving a loop-pile fabric that looked like a lamb's fleece when it was finished. It was used in the Turkish baths, for which that country was famous. Henry asked for a sample to add to the collection of memorabilia from his tour.

An idea was born.

Back in England, the Victorian population had just discovered the art of bathing, whether in a bath at home or in the sea while taking their holidays away from work. Henry's father William, back in the Droylsden district of Manchester, England, had a textile factory, and his brother was an accomplished engineer. Why not take this fabric back to England and encourage his brother to develop a weaving machine, or loom, that could make this fabric? Why not go into mass production and sell these very absorbent pieces of fabric to the Victorian population?

Henry's brother worked on the project and developed a weaving machine that carried two "beams" of yarn. The top beam moved faster than the lower beam and so created the loop pile, which in turn created the absorbency. This was the world's first towel, as we now know it, that was not hand-woven.

In 1851 the Great Exhibition was held at the Crystal Palace in London, and Christy's decided to exhibit their new towel. This exhibition was a celebration of all the great British innovations of the time.

Queen Victoria attended the exhibition and as she walked through saw the towels. She ordered six dozen for the Royal household and so was born the "Royal Turkish Towel."

This story has all the classic steps of innovation:

1. The exploration and the opportunity (The grand tour and Victorian bathing)
2. The connection to the solution (Harem and the Turkish bath)
3. Engineering and development (The Droylsden factory)
4. Execution (The Great Exhibition and the "Royal Turkish" brand)

See how important the exploration was. Also, notice that although the element of "Victorian bathing" came late in the story, it would have been sitting in Henry's subconscious throughout the grand tour. He lived in a different world from his brother but the two of them were "networked."

Notice the connection of two totally different environments, the exotic harem and depressing Droylsden. Networking between different environments is fundamental to innovation.

Notice also the "product launch," the purchase by Queen Victoria and the creation of a brand. Where is *your* Great Exhibition?

The final challenge of the innovator is execution, and be aware that the best ideas don't always make it! You now need a reliable delivery system as well as a "wow" product. Out of every 3000 ideas only one makes it! Execution needs discipline and the organization must now become really tight. The execution work is done primarily by the doers with some developers, connectors, and creators staying involved to ensure integrity and provide advice and support.

This is where the operations and sales people take ownership, but remember, they must be involved in the earlier stages so that there are no surprises here. Production problems are eliminated during development.

The value proposition that the sales people will use is also initiated during development!

In the Christy business that I just described and where I became chief executive, my works manager would be challenged with new manufacturing techniques or new outsourcing needs early in the development phase. The sales manager would have been engaging our close customers at the earlier connecting phase. He was always careful to engage those customers who we knew to be proven early adopters and never those who would plagiarize our new product with the competition.

The new manufacturing techniques might require new equipment and new operator skills. If you are in a service environment this might mean new software skills.

You will notice that at this stage in the innovation cycle I refer to sales and production as a single entity. Infighting and finger-pointing have no part at this stage. These guys should be integrated. Their engagement in the earlier stages of the cycle and especially in the developmental phase is crucial for speed of delivery at this final stage.

When we talk about integrity and honesty in the same breath there is a tendency to think they are the same. In truth, integrity is a prerequisite for honesty. Integrity means that parts are connected and function as a whole. This is the circumstance when you have the self-confidence to be honest. Why am I focusing on integrity and honesty at the execution stage?

The same is true in an organization. When you have an organization in which marketing, design, production, and sales all connect well, then you have an integrated organization that is honest and effective. This is vital at each stage in the innovation process and not least at this last stage.

We all have our challenges and mine was my chairman. I believed in having a delivery system that our sales people could rely on. He believed that if you constantly "wow" the customer, they will forget your previous failures to deliver.

Execution is payback time. If you have invested in a management system that has integrity, it will give you good data and information. If you have good information flows between the different areas of your business, and if you have integrity and honesty, you will beat the competition. Do not doubt that there is probably someone, maybe on the other side of the planet, who has thought up a solution to the opportunity you identified. Their solution may not be as good as yours but if you don't move fast they will beat you. They will even copy your superior product once it's out in the market.

At this last stage, again, time is of the essence. You will also notice that the loop has closed back to the sales people and also to the marketing people, who should be constantly seeking new opportunities.

At this last stage, you should know already who will be your early adopters. These are your first customers. You found them back when you where looking for opportunities. When it comes to selling down the delivery chain, you should have addressed the obstacles at the tipping point when you did the risk assessment. Too often it is assumed that selling down the chain is just another selling job. This is a new product; if you have not prepped well it could be a major failure at major cost.

The challenge for the sales people is to develop a value proposition and also collateral for the product. Supporting literature needs to be in hard copy as well as electronic format.

VALUE PROPOSITION

To place your product successfully with the customer you need to specify your *customer value proposition*. This has become one of the most widely used business terms. However, many value propositions claim benefit to the customer but lack evidence. You need to know where you perform the same as the competition, where you are better or worse than the competition, and where you and the customer may disagree. You then concentrate on the two or three positive points of difference.[1]

Ultimately, the customer will adopt the new idea if they see that the new idea makes their life easier or more enjoyable. These are intangible benefits, and the key word is "benefit." We have spent so long working on features of the product through both solutioning and development that we can forget that the customer is interested in benefits and not features.

The customer is interested in finding information quickly, not in the wonderful algorithms of Google. A customer is interested in paying less money when they fill their gas tank and far less on the wonderful fuel-efficient engine. The value statement has to focus on the customer's pain.

The value proposition has to focus on the pain and the customer's feelings. It needs to focus on two or three key points that will lodge in the customer's mind and one point of similarity with existing solutions. This way the customer connects to the existing solution and its shortcomings. They then read on to two or at the most three key benefits with the new solution.

The temptation is to list all benefits and create a shopping list as a catch-all for all customers. This approach has the major shortcoming that the list contains benefits that have marginal value to the customer and will create sales objections, cloud the offering to the customer, and obscure the primary benefits. This is a lazy solution and avoids trying to understand the customer's pain.

THE STATEMENT OF VALUE

Put yourself in the shoes of the buyer. You want a statement in the customer's language, and the opening statement needs to link you to a previous experience:

1. "Last time you filled your own gas tank it cost you $60; next time it will cost you only $30 to drive the same mileage."
2. "When you search competitors' Web sites for information, you can make two, or even three, attempts with a keyword and still not find the information you want. This software will ensure that a search on *your* Web site takes the customer to your solution the first time they search."
3. "Last time you cleaned the kitchen floor it probably took you 40 minutes with a lot of messy mopping and buckets. This mop will do it in five minutes and you need no other equipment."

Statements like these grab your attention. You are interested and you want to know more. The value proposition needs to link to the existing solution, with a "similarity statement" so that people feel secure. They will be cautious of having to change their behavior. Remember the previous chapter and the resistance to change.

First you need a statement like: "You can still go to regular gas stations" (this addresses concerns like buying propane), "You will use your existing software and it will not be impacted" (this addresses concerns about new software damaging existing software), or "You can store the mop easily" (this gets the user thinking in the context of existing cleaning equipment).

Then you need the two or three primary benefits the buyer will get from your product. A busy person's memory span is two or at the most three items and, remember, in selling business to business they are probably going to have to agree on the benefit with someone else before they buy your product. Make the sharing easy for them. List the two or three benefits in each of the two or three categories of time, money, and people. Then develop your master list.

DEVELOPING THE STATEMENT OF VALUE

Let's change from the everyday examples I have used so far and imagine a radical new training package that is being offered to your company. The features of this package are:

1. A workbook with "in company" examples
2. A supporting DVD with real-life examples
3. Training of trainers in-house to deliver training
4. Opportunity to break training from one day to four hours and two hour modules
5. A workbook in color with photographs
6. Discount for payment for training ahead of time
7. Evaluation of training effectiveness
8. Development of training delivery plan
9. All trainees are certified
10. Follow-up after three months to address any competency gaps

A general value proposition will probably capture the features that appear to provide benefits in the eyes of the training provider. Clearly a specific value proposition has to be developed for a particular client.

Starting with the general proposition and where the benefits are most likely, we might find that the attractive features from the supplier's perspective are the DVD, the color workbook, and the certified trainers, because these are the most visible to the suppliers of the training. However, for the company buying the training, the "in company" examples, the provision of training modules, and the three-month follow-up are probably what provide the best value.

Testing features to establish key benefits is something that should start during the development phase, and as a result the value proposition, with its "pain statement," the similarity statement, and the two or three key benefits, comes together easily at this final stage.

In developing the value proposition we also need to focus it on the risk points in the delivery chain. Remember Michelin's run-flat tire, first described in Chapter 8. They had two main obstacles or high-risk points: repair shops and auto manufacturers. The value proposition needed to focus here as well as at the consumer. The consumer ultimately pays for the innovation, but repair shops and auto manufacturers must be persuaded to install the technology. Look at the high-risk points on your delivery chain and develop your value proposition to fit.

NEW PRODUCT INTRODUCTION

In all the excitement of the marketplace it is easy to overlook operations issues. We are asking a group of people who produce the product or the service that ultimately creates value for the customer and revenue for ourselves to do something they have not done before. Did someone say resistance to change? Of course! If I am being asked to manufacture something entirely new and my production rate suffers and my income suffers, of course I resist.

Do you have that term NPI, *new product introduction*? Do you have a procedure for it? The product may be different every time but the procedure is the same, and yet you keep reinventing the wheel. Figure 10.1 is an example of what such a procedure might look like.

The pilot run of the new product should be initiated during the development stage to ensure that when you "go live" there is a minimum of disruption in the business.

Title page

Procedure title: Pilot test

Purpose:
To ensure that a new model or modified design can be manufactured to meet product specifications under controlled conditions.

Scope:
All new models or modified designs

Responsibility: Project engineer
Operations manager
Assembly supervisor
Quality inspector
Auditor

Definitions: SEI; Supplementing engineering instructions
Routing sheet; Description of equipment features

References: Routing sheet
Supplementary engineering instructions 2G-G04-09
Demo test report
Demo test operating procedure 2G-G09-03
Development schedule
Parts list

Records: Pilot and demo test report (retention one year)

Figure 10.1 Pilot test procedure.

1. The project engineer delivers the approved demo test report to the operations manager.
2. The operations manager ensures that a standard cost has been established and a quality plan has been implemented.
3. The operations manager finalizes the routing sheet and forwards it with the demo test report to the assembly supervisor.
4. Any difficulties in meeting lead time are advised by the operatons manager to the project engineer to enable revision of the development schedule.
5. The necessary parts for the pilot test are ordered by the project engineer on the finalized parts list from stores. A minimum of three models are manufactured.
6. The storekeeper delivers the required parts to the assembly supervisor.
7. New parts are approved by the project engineer. The supervisor approves the other parts.
8. The industrial engineer revises work methods and reassigns employees as necessary and advises the time allowed for operatons to the assembly supervisor.
9. The industrial engineering technician ensures that equipment, tools, and templates are installed and operational.
10. The order for manufacture is issued by the operations manager to the assembly supervisor on the development schedule as to date and quantity.
11. Manufacture is initiated and supervised by the assembly supervisor.
12. The pilot test is executed by following regular production procedures.
13. The quality inspector ensures that equipment is clearly identified during manufacture.
14. The quality inspector records any problems experienced during the test.
15. The equipment must be entirely manufactured on the production line. When manufacture is complete the equipment is audited by the project engineer, the quality inspector, the auditor, and any additional resource person required.
16. The auditor and a draftsman check the UPC number of the completed equipment in the GEAPR system and record these numbers in the pilot and demo test report.
17. Any problems experienced and responsibilities for action are described on a signed pilot demo test report, which is issued by the project engineer and filed by the quality inspector in the test file.
18. Copies of the report are issued by the project engineer to resource persons.
19. Noncompliances are corrected by the assembly supervisor before the equipment is shipped.
20. The project engineer ensures that corrective actions are complete.
21. Pilot equipment is released for sale on the approval of project engineering, the quality inspector, and the auditor.

Figure 10.1 Pilot test procedure. *(Continued)*

Doing it is getting the product to market		Strongly agree	Agree	Disagree	Strongly disagree
1	We get new products to market quickly				
2	We get the ROI we want on new products				
3	Few competitors are able to copy our products				
4	We penetrate all market channels and regions with new products				
5	We withdraw products that fail				
		×4	×3	×2	×1

Figure 10.2 Innovation process assessment.

Now score your organization on how well it performs on this final stage of the innovation process (see Figure 10.2).

YOUR INNOVATION PROCESS CAPABILITY

The maximum score in each stage of the innovation cycle is 20, and adding your five scores gives you an overall percent score. If you scored less than 50 then your organization is not in the game. If you scored over 60 but one of your areas is below 10 points then that area needs serious attention. Look at the activities I have described in that area and identify where you need to act.

Now you have an understanding of innovation strategy and process. Let's look at the human aspects of innovation in more detail.

BROWSER'S BRIEFING CHAPTER 10

- The best ideas don't always make it.
- A fast and reliable delivery system is essential.
- Engaging operations and sales people in the development stage prepares them for the execution stage.
- Competitors with inferior offerings will copy you if you are not fast to market.
- The customer value proposition should be confined to two or three key points.
- It is tempting to describe the features that have been the focus of the development stage.
- The customer is only interested in benefits.
- The benefit statement should link to the customer's previous experience and then show two or three advantages of your new product.
- It may be necessary to develop a different value proposition for each buying point in the delivery chain.
- The organization must have a highly specific NPI (new product introduction) procedure that was created during development.

Part III

The People

11
The Culture

He who innovates will have for his enemies all those who are well off under the existing order of things, and only lukewarm supporters in those who might be better off under the new.

—Niccolò Machiavelli

The chilling words of Machiavelli are a reminder that we have to plan very carefully the change to an innovative culture. There will be many who resist either overtly or covertly.

I am a student of culture. It fascinates me. I lecture each year at the University of Guelph on the subject.

The world of ISO gives me the privilege of seeing many cultures around the world. The subtle differences between the cultures of different nations never fail to intrigue me. Cultures are very much the result of a nation's history. Company culture is the result of a company's history.

The three cultures I know best are America, Britain, and Canada. I have learned a lot about the Latin cultures of France, Spain, and Mexico, and I am fascinated by the Eastern cultures of China, India, and Japan, especially as reflected in their art.

America's century of success as an innovator has come as a result of its diversity, driven by the famous words from the Statue of Liberty, "Give me your tired, your poor, your huddled masses yearning to breathe free, the wretched refuse of your teeming shore."

Britain, on the other hand, does not handle diversity well. A thousand years of repelling the invasion means the British treat foreigners with skepticism. On the other hand, Britain has always networked well, whether in its pubs or in its educated elite. Again, it has a strong history of innovation.

Canada is interesting. It has the diversity of the U.S.A., which gives it a culture of tremendous creativity, with a population only the size of California. However, unlike the U.S.A., it often fails in execution. It does not switch well from loose to tight.

We all struggle with what exactly we mean by "company culture." In Wikipedia it is defined as the:

i. attitudes,

ii. experiences,

iii. beliefs, and

iv. values

shared by the people inside and the stakeholders outside the organization.[1]

Deal and Kennedy describe corporate culture as "the way things get done around here."[2]

You hear the terms *strong culture* and *weak culture*, and there is a tendency to think strong means good. Strong actually means cohesive, but if this produces alignment with the wrong values it will not produce innovative outcomes for your organization. The strength may not produce innovation. *Values* are the beliefs and ideas that people see are important. A weak culture is of course one where the values of people vary and things only get done through procedures.

Both strong and weak cultures have downsides. The strong culture may suffer from groupthink, where a comfortable consensus overrides the tough decision. Nobody is prepared to challenge the status quo. The lack of challenge outside of an immediate task means innovation is a nonstarter. On the other hand the bureaucracy of a weak culture kills innovative thinking at the outset.

There are many types of culture; Deal and Kennedy describe, among others:

The Macho Culture with quick feedback and high rewards. You think of the stock market and sports teams.

The Work Hard/Play Hard Culture, which has replaced the Macho culture in terms of "fashion." This culture is the customer service culture with its own private language and focus on team. It is seriously risk-averse. This is the way of the "organization of the '90s" and does not lend itself to innovation.

The Process Culture, in which people are task oriented and information flow is poor. They produce consistent results, but always the same, and innovation is again a nonstarter.[3]

What you will see is that the culture of your own organization is a mix of behaviors and beliefs that have made things work for you in the past.

We all know the old saying, "If you do what you always did, you will get what you always got." If we are not an innovative organization at present then clearly something has to change. Unfortunately, cultures are very subtle and change is not easy. We have to ask, "What are the beliefs and values that must change?"

The sadness of much of quality management is that it has lost the focus on the customer and too frequently has focused on internal efficiency, using the customer as the excuse for driving that efficiency. This is short-term thinking and in truth focuses on the short-term interests of the shareholder or business owner and not the long-term interests of the organization.

Long-term thinking is interested in what today's customer needs tomorrow, and who will be a new customer tomorrow.

Quality management has also focused too much on process and procedure and explicit, or documented, knowledge. The beliefs and values in an innovative organization will need to shift to a better balance of people and process and a greater emphasis on tacit knowledge.

There are a number of values and hence behaviors that an innovative organization must develop. They all stem from the need to develop a learning organization.

BEHAVIORS ARE BASED ON VALUES

The behaviors that most organizations need to develop in order to become more innovative are:

1. Exploration
2. Interaction
3. Observation and note-taking
4. Collaboration
5. Experimenting
6. Embracing failure

EXPLORATION

In today's business world, exploration is the victim of "good time management." We have become obsessive about process efficiency in our operations and have extended this mind-set to all aspects of our business life. Archimedes's moment of "Eureka!" was when he was having a bath. Your best ideas will come when you allow them instead of forcing them. We must allow time to "bathe" and relax. Google is an organization that does this well.

Allow yourself, and your teams, to step out of the box and explore. Leadership teams have retreats. If we are going to release the collective subconscious, *everyone* must step out of the box.

Some of my best ideas ever have come from chatting at the bar in rugby clubs. I do my best writing in coffee shops while I am people-watching. I choose to live out of the city so that I can release the learning that has accumulated during the interactions of the day in the city.

My innovation workshops are at a beautiful location called Kingbridge. Everyone loves their day out of the box. They feel the ideas coming free.

Go to places you haven't visited before, meet people you haven't met before, experience things you find a bit intimidating. Remember the old saying "travel broadens the mind." You should join a network. There are many around these days. I don't mean Web-based, I mean face to face.

I am fortunate to be deeply involved in ISO and that gets me traveling many places. I always take an extra couple of days after my meeting to experience the culture, whether it is the street market in Nairobi, a police chase in Kuala Lumpur, a mummy's tomb in Egypt, or Dracula's castle in Transylvania. You don't have to travel the world by the way. The world is in your own back yard! I recently drove across Iowa and developed a large chunk of this chapter at the Texas Grill in Des Moines. New experiences broaden your perspective. They also enable you to see new opportunities. You get these opportunities to travel. Take them. Then allow yourself time to step back from your experiences. I have that fortune cookie note that should be the mantra for every innovator: "The secret of a good opportunity is recognizing it."

INTERACTION

Mix with interesting people. This may surprise you, but I do not mix easily. Put me in front of a thousand people to share my life with them and I have no problem. Put me in a room of fifty garrulous and assertive people at a cocktail party and I go quiet.

Let me give you a tip. If you want to find someone interesting to talk to at a cocktail party, a great technique is to relax by a table, visualize your favorite room at home and that the people in the room are all your best friends. In five minutes this visualization will draw to you, almost like a magnet, someone in the room who thinks you look like an interesting person.

A great book on networking technique that I can recommend to you is *Never Eat Alone* by Keith Ferrazzi.[4]

OBSERVATION AND NOTE-TAKING

Always have a small notebook with you. Have a book that fits in your pocket and is a book that you are proud of. I bought mine at a store called Indigo. It has a leather cover with beautiful embossing. The pages are perforated and will tear out without damaging the book. Writing in the book is a wonderful experience for me.

However, life doesn't always work like that. You may get caught without your book. I was sitting in Charlie Biddle's jazz bar in Montreal listening to Charlie himself on the bass. I had been struggling for months to find an opening chapter for my first book. My thoughts suddenly came together in my mind and I had no paper. The lady at the bar gave me a white paper placemat and a cheap pen. My opening pages of *Do It Right the Second Time* were written on a paper placemat.

Charlie finished his set and, walking past me, thought I was writing a song. I won't forget the words of the late Charlie Biddle. He said, "Write one for me, man."

Make a point of traveling in different cultures to get new experiences. Innovators always take notes. You should always be learning.

Observe with an open mind and ask why things happen. Don't judge. Find out about other people's lives. Too much of business training is analytical and judgmental. Write notes later but not too late. The ancient practice of keeping a diary is exactly this. Always be looking for what bugs people, but do not try to solve their problem; simply find out what their problem is. Solutions come later. And if you want the real truth, talk to kids.

One final piece of advice—always travel on Robert Frost's road less traveled.

COLLABORATION

Find the three or four other people that share a passion for your topic. This is networking but more focused. This is the beginning of a *community of innovation,* which I describe in Chapter 12. I collaborate with people in

Europe as well as in North America. Tools like Skype make this so much easier these days.

Freely exchange your knowledge. When others send you articles or ideas, take time and space to read them and see what ideas they trigger for you. Every six to eight weeks I do a Skype call with my friend Chris Hakes in Europe. The interaction always generates fresh perspectives. In between we fire off articles to each other that we think will be interesting.

EXPERIMENTING

Experimenting means being prepared to fail. It also means being surprised with an unexpected outcome. Post-It notes came from a failed experiment, and the name WD-40 refers to the 39 previous failures.

Prototyping needs to stay at the concept stage. Use techniques like storyboarding that are tactile and interactive. Use tools like . . . yes, Post-It notes, that again are tactile and interactive.

When I teach process mapping I always use paper and Post-Its. I refuse to use the computer screen. It is too controlling and excludes 80 percent of the participants. On the other hand, use the technology of video extensively. This is a wonderful learning tool. Your culture must accept unpolished presentations. Polished PowerPoint presentations exclude too much of the original thinking.

One other key point: involve your customers, managers, and business partners continuously in the experimenting, and always have multiple prototypes to show.

EMBRACING FAILURE

We find this hard to do. The "take care of number one" mentality, which frequently broadens into a corporate culture of mistrust and backstabbing, leads to people being inherently skillful at distancing themselves from failure. People see a project failing and they stop attending meetings and stop returning phone calls. A project finally fails and there is no project closure or lessons learned session.

It's like not doing an analysis on your customer complaints. These are one of the richest sources of knowledge.

One of the commonest failings I see in project-driven businesses is the absence of project closeout discussion, analysis, and documentation. The excuses are always "we ran out of time" and "people got moved to other projects." You must learn from experience.

I am now going to add a seventh behavior to the list.

RECOGNITION OF BEHAVIOR

You must recognize new behaviors, not results. Recognize innovations and not heroic defenses of the old. Endorse the behaviors I have described. Reinforcing new behaviors builds trust, especially when they are based on the new values that they imply. One other point—you must move eventually to a point where recognition and reward are aligned. Phil Crosby, in the 12th of his 14 steps, said, "Appreciate those who participate." I talk more about recognition in *Do It Right the Second Time.*[5]

CHANGING CULTURE

So how do we begin this change of culture?

The challenge is clearly one of *Leading Change,* which is the title of John Kotter's book on the subject.[6]

There are a number of issues that will create resistance to the change to an innovative culture. You see these from the new behaviors I have described and also from looking at your existing culture. The primary change agents will be your "I" team, and I will describe the change process more fully in Chapter 19, "Lighting the Fire." Resistance will be especially strong if you have a history of success. It will be necessary to create a sense of the need to change. The process for handling change should follow this approach:

Create a sense of urgency. Identify a falling revenue item or market share with a key product or service or customer. Pick an item or a customer who is high-profile and develop a sufficient desire for action. This is Kotter's "burning platform."

The "I" team. A change agent team must include people outside management and must be designed carefully. Their mission is to quickly create a critical mass of believers in the organization by replacing a dying product. This new product is the first visible evidence of developing an innovative culture.

Early win. The "low-hanging fruit" is captured. This is a product or service or customer opportunity with high benefit. Wins must be created and not just based on hope. The win must produce a quick result and then be publicized. You use a rapid version of the innovation process described in Chapter 19 to achieve this.

Communicate vision. The "I" team must draw in the people by focusing on the two or three most important successes from the early win. These successes must be easily communicated in concise and simple wording. The team must keep repeating the successes. A "One Minute Message" should be developed. Communicating the vision becomes a key task for the business leader.

Enable action. People must be given the authority and time to innovate. Leaders must find the obstacles and remove them. Compensation must be structured to support innovation.

Don't declare victory. The short-term win creates danger as people relax. The outcome must be monitored and negative side effects dealt with. Significant culture change must be ensured before moving on to bigger issues.

Anchor new approach. Fundamental culture change is essential until innovation becomes business as usual. Culturally, the next generation must personify the innovation generation. Innovations must be explained and the participants recognized.

Recognize success. The behaviors described earlier are endorsed. Building trust reinforces new behaviors when they are based on the new values. Recognition and reward are aligned.

This gives you an outline of the change process; the business leader plays a key role in all of this.

MAINTAINING AND STRENGTHENING VALUES

Whether you call it performance management or annual assessment, the one-on-one performance review is one of the key activities that will either endorse or undermine the culture you want. If you want to be innovative, the goals that people first agree to and are then assessed on need to reflect the values and behaviors of innovation. This is where you take time to see how much exploring and note-taking got done. Who were the collaborators, which experiments succeeded, and which ones failed? This is where you congratulate people for having the courage to fail. If these behaviors are listed in the review input and again in the half-year assessment, they will start to be taken seriously. This will also develop the other key components of an innovative culture, trust and honesty.

RECRUITMENT

The other process in which you can unknowingly undermine your culture is your recruitment process. An innovative organization embraces diversity. There is a tendency for us to recruit more of the same when we look for new people to join our organization, and we mirror the majority of our company population. Your recruitment focus should be on the values that matter in your culture. You must evaluate whether potential new people support those values and practice the behaviors you seek.

When I joined the research division of Courtauld's, the company had each potential employee spend three days in interviews. This was the way in which they ensured that new people had the right values. Compare this investment with the cost of labor turnover. Companies are often surprised that when they invest in only a one-hour interview they also have high labor turnover.

VISION

When you know what values you want, a key task for the leader is the development and communication of a vision that embraces these values.

Research In Motion (RIM) is a company that I have had the privilege of working with as they developed their management system. Their CEO, Mike Lazaridis, is the visionary who has created and communicated the vision of innovation in that company. You can feel the excitement and dedication within the walls of RIM as their people go that extra mile.

IBM is another company I had the privilege of working with as Lou Gerstner took them through their low point in 1992 and then transformed them from a company that "made computers" into a consulting firm. Their business is now about knowledge and not hardware. Gerstner was a great visionary, and I remember the great surprise in the market when IBM, a great innovator, purchased Lotus, one of the early steps in that transformation. Gerstner didn't just innovate their product but their entire business model.

Finally, Steve Jobs of Apple is an example of an innovative leader. Look at the Apple share price in Figure 11.1.

The leader's task is to communicate a vision based on the values that encourage the new behaviors I have described.

The values and behaviors I have described manifest themselves in the small groups of people that come together at the start of the innovation process. These groups are referred to as *communities of innovation* or sometimes as *skunk works.*

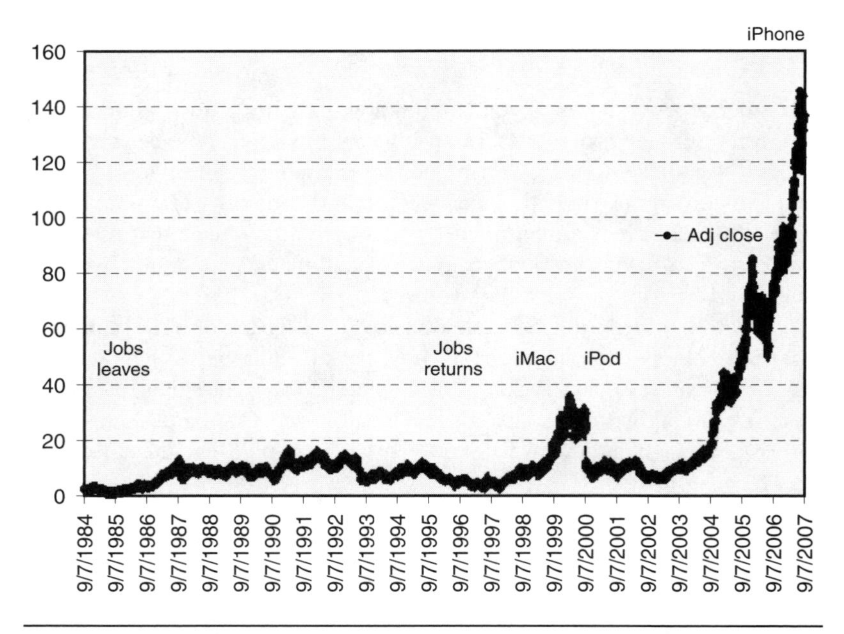

Figure 11.1 Apple share price before and after Jobs.

BROWSER'S BRIEFING CHAPTER 11

- Culture is "the way things get done around here."
- Strong culture means cohesive; weak cultures have variation.
- Culture is based on values or what is important.
- Cultures such as "macho," "work hard, play hard," and the "process" culture do not support innovation.
- An innovative culture has to be strong to be effective but must be based on unique values.
- Behavior is based on values; innovative behaviors are exploration, interaction, observation, note-taking, collaboration, and experimentation.
- Embracing failure is a critical behavior.

Continued

Continued

- Recognition endorses values.
- Values are maintained and strengthened through the performance review process and through a sound recruitment process.
- In order to change to an innovative culture, an organization will need to go through a distinct change process.
- The business leader has a key role in communicating values through a vision of innovation.

12
Skunk Works

No man is wise enough by himself.

—Plautus

Since I wrote the chapter on teamwork in my book *Do It Right the Second Time* much more has been thought, said, and written about the challenge of working in groups. In that first book of mine I talked about the teamwork in the game of rugby, and I talked about understanding my own strengths and limits as a team leader. A rugby team is, in fact, two teams: the *forwards* or "scrum," which are eight "big guys," and the *backs* or "three quarters," who are the six "fast guys." These two teams are joined by a cheeky little guy called the "scrum half." The scrum half is the critical link between the forwards and the backs and spends much of the game constantly chattering to both backs and forwards.

In *Do It Right the Second Time* I talked about how as captain I gave the leading of the forwards to a bearded mountain named Andy Alltree and the leading of the backs to an Irish wizard named Brendan Webb. I played on the wing because speed was my asset and it gave me the chance to read the whole game, view the opposition's strategy, and guide the team if we were losing our direction.

You can see how these roles play out in your own innovation team and how this manifests what I call "de Bono's law," that the natural span of communication is six to eight people. A lot has been written about the experience of running teams of twenty to a hundred people but I will come to that later.

In networks, information becomes knowledge, and the best innovation comes from organizational knowledge. It is the communities of innovation that I mentioned earlier that enable the best knowledge transfer.

The coordinator of this community must have a passion for people, provide resources and tools, and also introduce new members. The coordinator monitors the big picture, such as number of e-mails and level of participation, and identifies the issues to be addressed.

Your role in the community will be as a creator, connector, developer, or doer. The blend of these four attributes will shift as you move through the innovation process. At the first stage of the process there will be a majority of creators. At the last stage a majority of doers.

The famous solution to this problem of bringing people together to innovate is a "skunk works." Lockheed in the 1940s created the term when they named their community of innovation after the moonshine distillery in the cartoon strip Li'l Abner. *Skunk works* is now a widely used term. Skunk works often operate without the knowledge of senior management.

Communities of innovation are a fancy name for skunk works, and this term is the product of knowledge management. They are multifunctional networks of relationships and don't look like teams. Trust is essential, and they recognize that learning is a social activity. The volunteers have a passion for the subject.

As I write, I have just returned from chairing a meeting in Rio of the international working group (WG) developing the standard ISO 10018, *People involvement in management systems.* There were some critical tasks to be performed in our meeting as we created the first working draft of the standard, and we had just four days to do it. My own task was to make sure we got to "end of job." Our finished document had over 12,000 words and, amazingly, Anne Wilcock from Canada captured the input on her laptop. Her role titled "secretary" was really the huge task of information manager. Research has shown that too much shared information becomes a problem for a group over time. We broke the week into 90-minute segments, and Paul Simpson from the UK acted as timekeeper and kept us on track the whole week. He was time manager.

We had a good sense before the meeting of where people's aptitudes lay. Anne had documented the work of the preceding study group and Renato Lee from Brazil had been language monitor when he had worked on an earlier standard.

We were a diverse group with different native languages but our working language was English. Renato ensured that everyone understood the terms that were used in conversation. Lille Harnell from Sweden kept us honest and challenged the team if we were galloping ahead and missing something important. All of these roles were shared and there was overlap. A good team does not compartmentalize the roles of its members. At the same time, people take responsibility for their own roles.

PASSION

The "people group" I described comprises people who have experienced firsthand the problems from not involving people in the running of a business. They have a passion for the subject, and rather like in the game of rugby it is a game of passion and friendship and all newcomers are welcome.

When you create your community of innovation you need to know in advance who are the creators, connectors, developers, and doers. Ensure that you have a good mix, and of course an overriding attribute is a passion for your subject.

In forming a group we must be clear on its purpose and we must also be clear on the type of contribution each person can make. We may give a person a role but they may demonstrate strength in another area over time.

In the early days of your community you will operate in a loose mode, and there must be a predominance of creators who will see opportunities. The creators will network with others both inside and outside the organization. Remember Linus Pauling. "The best way to get a good idea (opportunity) is to get lots of ideas (opportunities)." The techniques of brainstorming I described in Chapter 6 are what you use to develop these opportunities.

EARLY RESULTS

In these early stages the leader has an interesting challenge. You would think that the challenge is to develop the network and the relationships so that people work well together. In truth there is the bigger challenge of showing results. An early win is vital. The leader has to be process-focused and ensure that there are documented outputs even at the earliest stages. If not, the team will lose its passion and rapidly disperse.

Still in loose mode the team will start to capture some potential solutions. The task of the people group that I mentioned was the creation of a first working draft (WD1). This WD1 listed problems they had encountered through lack of people involvement in management systems. Inevitably we would find potential solutions but we would not discard those solutions. We would record them.

There will be a point, though, where the team morphs into *solutioning* or *connecting* mode. This is where the team numbers are likely to balloon, and where from a team perspective new challenges arise.

COLLABORATION

Work by Gratton and Erikson confirms our own empirical experiences.[1] As teams grow beyond the 20-person mark, and also as teams become more virtual, collaboration decreases unless collaboration is structured into the work. Complex teams are also less likely to help one another.

The factors that inhibit collaboration are nationality, age, education, and especially length of time on the team. In the world of ISO standards I see the first three issues addressed very positively and they have become a great asset in creating standards that are accepted internationally. However, I often see a group that started a particular piece of work feeling threatened as new people join the group. It is important to remind ourselves that more people means more ideas and more solutions.

One other thing to beware of: the greater the proportion of experts on a team, the greater the likelihood for conflict and stalemate.

There are a number of actions the leader or facilitator needs to take to overcome these problems. They are all actions you are familiar with, but let's list them.

A LEADER'S ROLE

First, face-to-face meetings are imperative. It's not just about transferring information, it's also about building relationships. Trust is based on behavior and being able to rely on others. This only truly develops in a face-to-face situation. Don't slip into the conference call mentality.

Secondly, new people must be mentored into the group, not through a formal HR procedure but through informal and personal attention from the leader. A one-on-one meeting or a personal phone conversation if geography is a problem should be held prior to the main meeting.

During the main meeting it is essential to talk off-line and ensure that the new people are truly engaged. The gift of time is one of the most precious gifts from any leader and is truly appreciated.

Thirdly, and this will be surprising, research has shown that in the early stages of the team, roles need to be well defined. While at the same time the approach should be loose. Many think that leaving roles vague encourages sharing whereas in practice, having a defined role makes people feel relevant and causes people to feel that they can work independently and contribute to the group outside of meetings. Another research finding is that in the creative mode, having a loose process leads to people investing more time and energy in collaboration.

You can see that a number of these actions seem to work against each other and so the job of a leader or facilitator is complex and requires balance.

When the group reaches its tipping point it also reaches the toughest job of all. The mode of operation changes from loose to tight.

CHANGING FROM LOOSE TO TIGHT

This change is not sudden. The group should prepare for it. They will have been looking at their list of potential solutions and assessing risk and usability. The list is condensed down to, say, six or at the most ten preferred options.

The tipping point is where the group reconnects with business strategy and two or at the most three "hot options" are selected. I talked about this in Chapter 8.

Moving forward from here, the team changes. Some of the highly skilled creators and connectors will leave. They go back to looking at new opportunities. New faces will have joined at the solutioning stage. Depending on the numbers you have, you may need more developers to join the group at this point.

If you are a small or startup business your challenge is much greater. Your team is going to comprise the same people and so the behavior change in the team becomes much harder. Conversely, the project management for a small business is much easier.

THE SHIFT IN TEAM CONTENT AND CULTURE

The challenge for innovation is that we need two cultures. Many of the behaviors are the same in the two cultures, but there is a fundamental difference between stages one and two, where there is a need for freedom of thought, and stages four and five, where there is a bias for action. Stage three is the tipping point.

Many organizations have solved this problem by separating the two cultures into two distinct parts of the organization. This seriously impedes the communication down the innovation chain. Far better to transition the culture between each stage of the process. See Figure 12.1.

You can do this by knowing the kind of people you have and grouping them at each stage in the process so as to encourage the behavior you seek.

Find the opportunity	Connect to the solution	Make it user-friendly	Get to market
Creators			
Connectors			
Developers			
Doers			

Figure 12.1 The shift in team content and culture.

At the development stage speed becomes important so you will probably break the group into a series of task groups each addressing distinct aspects of the solution. The solution is now starting to become a new product or new service, and each of these task groups will address a component or element of the product.

HANGING TIGHT

At the development stage the challenge is to develop usability of the new product, and the group will engage customers to assess whether the behavior change required by the product or new process is too extreme.

Your picture of the group should now be shifting. Instead of an amorphous, free-flowing mass, it has now segmented into a set of tight fighting units. Each unit probably has only five to seven people.

Each of these fighting units will be highly task-oriented. They will have very specific goals. The goals will be SMART or should I say "SCHMART." That is, *specific, challenging, measurable, agreed, realistic,* and *time bound*. This may not be new to you. One task may be sourcing critical suppliers or subcontractors and setting up contracts. Another will be refining a component or subprocess. A third will be testing alternatives on the early adopter. The leader has to keep these task groups aware of each other's work. Frequent huddles are essential. Maybe even a couple of times a week or more.

The new product and its components are being tested and the overall group is working to a tight project plan. This is a familiar mode of operation for anyone familiar with an R&D environment. Where most R&D operations go wrong is that they are disconnected from the real world. A lot of serious questions are being asked about the effectiveness of R&D. IBM's global survey has shown that little more than 10 percent of new innovation is coming out of R&D facilities. A lot of this is due to their isolation and obsession with secrecy. It is essential at the development stage to be connected with both customers and suppliers, as well as your own operations and sales people.

The task groups at this stage should be keeping concise and understandable records. Communication between task groups is hard. Records should not be long dissertations. They should be crisp and easy-to-read notes. Networking between teams is also critical, and in a large group each task group should have a liaison person whose job is to scan the work of the other task groups.

At the final execution stage the team changes again. You will keep a creator and a connector and certainly a number of developers, but here is where operations and sales people become the major part of the team. These people are very task-oriented and so the team culture will shift once more. The sales people gain feedback from the marketplace and the operations people from manufacturing and service delivery. This enables the developers to further refine the product.

You have seen from this chapter and the previous chapter that a number of key competencies are required. In the next chapter we will look at the attributes of the competent innovator.

BROWSER'S BRIEFING CHAPTER 12

- The natural span of communication is six to eight people.
- Skunk works is a widely used term for communities of innovation.
- The coordinator of the community must have a passion for people, identify key issues, and provide resources.
- The community will start in loose mode but it is essential to achieve an early result.

Continued

Continued

- As a community grows in size, collaboration becomes more difficult.
- A high proportion of experts impedes collaboration.
- The leader or coordinator must ensure 1) face-to-face meetings, 2) mentoring of new members, and 3) clear role definition for team members.
- The community changes from a loose to tight mode of behavior after the tipping point.
- Team membership may shift after the tipping point.
- For a small business where membership can not change, behavior will need to change.
- In a larger organization, at the development stage, the community may break into task groups.
- The community must involve customers and suppliers at the development stage.
- Innovation succeeds through the sharing of group knowledge.

13
The Competent Innovator

I keep six honest serving-men
(They taught me all I knew);
Their names are What and Why and When
And How and Where and Who.

—Rudyard Kipling

When ISO 9000 changed in year 2000, there was a subtle shift in one of the requirements of ISO 9001 from "training" to "competence." The old standard had focused too much on the delivery of "training" by an organization and too little on the acquisition of knowledge, skill, and ability by the individual. In truth, most individuals acquire most of their competencies through that terrible thing called experience and not through formal training. I had a boss once who said, "The worst thing about experience is getting it."

Education and training provides the initial knowledge framework for developing competence, but the experiential learning that follows is where the skills and abilities are developed.

I have seen so many companies struggle with this. It is one thing to provide a formal education or training session, but a whole other thing to provide that essential mentoring that follows. The IT revolution of the '90s saw the removal of a whole middle layer of managers whose hidden task was the mentoring and guidance of their staff. Those that survived the revolution were often diverted into planning roles as business became more complex. As a result, organizations struggle with development of competence.

Nevertheless, everyone will first need an understanding of what innovation actually is. The chances are your own perception has changed since you started reading this book. A one-day workshop addressing the content of this book enables people to participate in discussion, address their concerns, and participate in exercises and assessments. A number of the skill sets described in this book can be introduced in such a workshop.

Your leadership must understand what innovation is. There are many myths and legends. They must understand the difference between creativity and innovation and that innovation is not magic and there is a process. Everyone has a role in the process, and the outcome will be achieved through collective knowledge and not the lone genius. Finally, the innovative outcome will not be some sudden epiphany but will be the result of following the process.

In educating the leadership I would encourage you to focus in the morning of the workshop on the individual and in the afternoon on the process and the organization. Engaging the individual and showing them the value they can personally provide is a big surprise to many. You hear people say, "Oh, I am not creative and so I can't innovate." Once they have done the self-assessment they realize what their contribution will be. You then develop the process and show them what they will actually be doing. The workshop should be structured in 90-minute segments with at least one exercise in each segment.

In the first segment I have a brief introduction explaining what innovation is and what it is not. I then ask each member of the group to answer two questions.

First, "What is your primary concern about your organization's ability to innovate?" and second, "What product or service causes your customers the most frustration or difficulty?" This gets everyone involved, and they are fairly easy questions to address.

I then introduce the group to the essentials of knowledge management, which we covered in Chapter 2 of this book. The difference between the workshop and reading this book is of course the interactivity and the development of collective knowledge.

In the second 90-minute segment the group carries out the self-assessment you saw in Chapter 3 and finds out whether they are creators, connectors, developers, or doers. They also do an exercise in creativity that emphasizes the importance of interacting with others. While on their own they may only find three or four solutions or ideas; through the process of structured brainstorming they can find 15 to 20 ideas.

Figure 13.1 shows the workshop I developed with Ken Bales at CSA when we educated the CSA staff on innovation.

The afternoon moves to an organizational level and the shift in organization structure that enables innovation to flow more easily. The group does an organizational assessment to identify where their individual organizational weaknesses lie.

1. *What is innovation?* Innovation is an imperative in order to compete in the marketplace. Innovation is driven by people's desire for convenience and comes from the use of new knowledge to create new products and new services. Conventional organizations, today's client, and fear of failure all inhibit the creation of new knowledge. You learn the critical factors that lead to the creation of new knowledge.

Exercise: Group assessment of the concerns and opportunities in their organization.

2. *How to get innovation.* You learn how Innovation builds from the knowledge created in a quality management system. This knowledge is fully realized when an organization moves beyond the process approach to a network structure. Building on a culture of knowledge management leads to a culture of innovation. The best innovation comes from the wisdom of crowds. Collective knowledge is captured through communities of innovation.

3. *How can I contribute to innovation?* The innovation roles are the creator, connector, developer, and doer. You identify where you will make your own best personal contribution to innovation, whether by generating ideas, finding solutions, making solutions practical, or implementing solutions.

Exercise: You will determine where you will make the best contribution to innovation.

4. *The innovative organization.* You learn that a different kind of organizational structure is needed to lead to the creation of breakthrough innovation. You reduce hierarchy while increasing autonomy and diversity.

Exercise: You evaluate the innovative strengths and weaknesses of your organization.

5. *Innovation strategy.* Integrating innovation with the strategic planning process of the organization is essential. However, this is often done in a manner that is ineffective. You learn how to develop customer-focused innovation and open market innovation in which you work with your customers and suppliers in developing new ideas.

Exercise: You sacrifice what is sacred in today's product and learn how to find that white space where there is no competition for tomorrow's product.

6. *Execution.* Do people truly want our new idea? Typically for every 3000 ideas generated only one makes it. How do we select the ideas into which we invest our resources? The key elements of the execution plan such as risk assessment and risk mitigation are explained. You learn the importance of evaluating the behavior change required in using a new product.

7. *The path forward.* The road map for innovation is summarized. The importance of the burning platform and the short-term win are explained. The method for obtaining the short-term win is outlined. These are first steps on the road to an innovative culture.

Figure 13.1 Agenda for one-day innovation workshop.

We then start to address strategy and how it starts with customer needs and then broadens to market need and open market innovation. The people in the group take the idea they listed in the morning workshop and address these questions:

- What do our least favorite customers want?
- What do customers hate about buying our product?
- What aspects of our existing product or service are sacred?

This way the organization starts to see its own "white space" opportunities.

The day concludes with the leadership team initiating its own priority plan for deploying innovation into their organization.

BEING CONSCIOUSLY INCOMPETENT

At this point we have reached the stage of being "consciously incompetent" and we must move to the next level of "conscious competence." This means that we must not just educate but also train people. Education provides knowledge. We need to be able to apply that knowledge in the form of skills in order to get results. You conducted an assessment of personal innovation aptitude back in Chapter 3. This is an initial guide, and people will be able to contribute in more than one area of the innovation process. However, you will need additional skills and tools. I explained how behavior is the basis of culture so you should be able to see how culture and competence align.

COMPETENCE

The word competence is cumbersome, but it is fashionable so I should use it. I much prefer the five-letter word "skill" but that word has come to be applied more in a physical environment than in a thinking environment. The terminologists in ISO 9000 do a good job on terms and definitions and their convener, my friend Bill Truscott, will be thrilled to hear me say that his team has given us a definition of "demonstrated ability to apply knowledge and skills." There is another definition from the training standard ISO 10015: competence is the "application of knowledge, skills and behaviors in performance."[1]

You can see the word "behavior" in the second definition. I will take you through some of the behaviors you should engage in, some of the things you should do in order to develop your skill and be a better innovator.

The kinds of skills you need to develop in each stage of the deployment are:

1. and 2. Find the opportunity (creators) and solution (connectors)—exploration, interaction, observation
3. Selection of solution—risk taking
4. Development of solution—experimenting, embracing failure, project management
5. Execution—value proposition, storytelling, strategic selling

Just like networking, note taking and collaboration apply right through the process, and even the skills I have indicated in a particular process stage for their primary deployment will be used in other stages of the process.

The primary attribute for the innovator is a "yearning for learning" so we must develop a thinking workforce that is constantly looking for a better way and is willing to get ideas from anywhere. This leads to other competencies such as an ability to interact with others and willingness to suspend your own judgment.

As I take you through the different competencies you will see that they overlap, so be ready for this. Taking these in turn:

EXPLORATION AND INTERACTION

This may not seem like a "competence" but more of a behavior, and yet in today's world we have a tough time doing this. We do have to train ourselves to explore. What you really need here is a plan both for yourself in your personal life and in your organization. However, be ready to break the plan! Let your emotion lead you and not the plan. I will explain more about this in Chapter 14.

Innovations happen at the intersection of disciplines. Dr. Niall Ferguson at Good Hope Hospital has told me how this has become fundamental in the medical profession. A good physician will not just use their own knowledge but integrate it with the knowledge of the patient and the knowledge of the patient's family. The problem may reside in one area of knowledge and the solution may reside in another. Innovation takes place in the space between people. The Internet allows people to share ideas but most breakthroughs come when you are face to face.

This skill is one you use when you interact with customers and suppliers. This is another skill you can develop on your own but can be used both at conferences and in day-to-day activities outside of work.

QUESTIONING AND LISTENING

Remember Lt. Columbo? He wasn't afraid to ask dumb questions. Every case was a learning experience. Columbo was an auditor at a crime scene. Good auditors are not predators, they are people who want to learn. Remember, the innovative organization is a learning organization. The innovative person is a learning person.

When I train auditors I always share the quote from Rudyard Kipling in his book *Kim*:

> I keep six honest serving-men
> (They taught me all I knew);
> Their names are What and Why and When
> And How and Where and Who.

Start your questions with any of these six words and you will learn. Your questions must go deeper though. You are searching for people's feelings, what frustrates them, angers them, bores them, or wastes their time.

Questioning without listening is a waste of time. As Stephen Covey says in his *Seven Habits of Highly Effective People*, "Seek to understand." This means understanding the feelings behind what people are saying.[2] If a person has a lot to say, ask if you can take notes so you won't forget what they said. Above everything, understand their joy and understand their pain.

Another technique to use here, which will be familiar to people in quality management, is the *five whys.* A very simple tool and a great way of getting to the heart of an opportunity. I remember sharing this technique with a group when I was delivering a continuous improvement training course. When I roleplayed the five whys one of the guys in the group said, "Just like my kids! Why daddy? Why daddy? Why daddy?" Kids are not afraid to ask questions. Follow their example.

NOTE TAKING AND TECHNOLOGY

If you have a notebook that you are proud of, then you will continue to take notes. Make your notes as cryptic or as lengthy as you wish. This has to be your own style.

The notebook you use can be leather bound, moleskin, or whatever, but let it have style and make it be something that you are proud to own. The paper inside is just as important. When paper is good your pen flows on the paper. That will inspire you to further thoughts. Some people like index cards because they are easy to organize. You can also buy index cards in the form of Post-It pads, which makes them easy to carry. I like Post-It notes with lines on them. Before my BlackBerry arrived I used to be "King of the Post-Its."

One other technique I am using more and more is that I take notes on my BlackBerry and then e-mail them to myself. When I return to my desk the notes are waiting for me on my PC and they are already typed. The other piece of technology I really like for note taking is mind map software. See Figure 13.2. I use Mind Jet but there are others for you to try.

I can't stress enough the importance of note taking in the innovator environment. You will often have an idea one moment and forget it the next. Look at the cartoons of Leonardo da Vinci. Those are the notes of a great artist and engineer. Note taking has the added bonus of "clearing the mind" so other thoughts can then emerge.

Your notes can be verbal or graphical. If you are graphically oriented, the mind mapping technique will probably appeal to you. If you have mind mapping software you will find it an excellent way of laying out your thoughts.

As you make notes, circle your keywords. Keywords are very important when you return later to read your thoughts. When you are back at your desk, use Google to search for more ideas on the keywords you have circled.

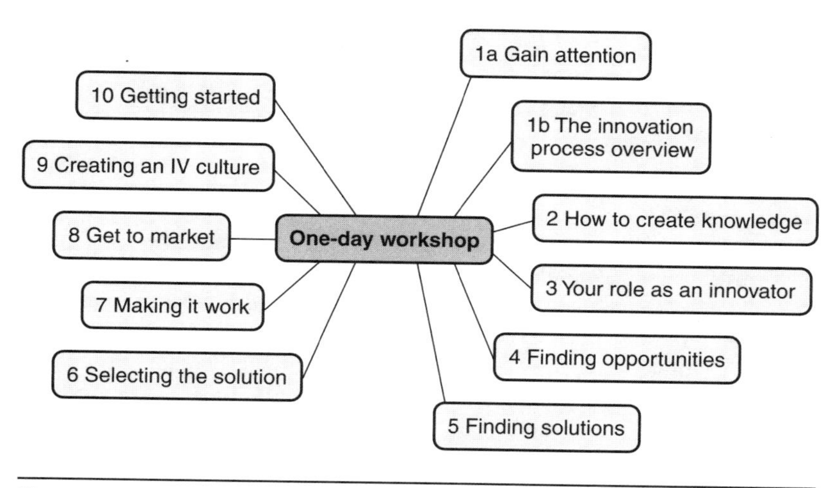

Figure 13.2 Mind map for an innovation process training session.

NETWORKING AND BRAINSTORMING

These two skills are very, very similar. I've talked about brainstorming earlier in this book, and in this chapter I've talked about the skills you should use when you are networking. Let me just focus here on networking attributes. They all center around my good friend "balance."

1. Balance with people who have something to give you and to whom you can give.
2. Initiate the giving, balance the input and output.
3. Balance work and nonwork interaction.
4. Balance your outgoing ideas with quick responses to incoming ideas.
5. Balance trust and respect.

I will talk specifically about networking in the next chapter.

I talked about brainstorming in Chapter 5 so won't dwell on it here. This is networking on a smaller scale. But remember, people with low social anxiety create more radical ideas. Don't expect a rare and occasional brainstorming session to generate innovation. It has to be practiced every day. The individual brain is not the source of creativity; it works much better interacting with other brains.

BEING PREPARED TO EXPERIMENT AND TO FAIL

This is the hardest skill of all. Remember WD-40, named after the 39 previous failures? That was a testament to the learning that came from those previous failures. We make mistakes every day and yet we keep repeating them. House buyers view homes on weekends not weekdays, and yet when we are selling a house we expect a buyer every day. When we travel we take too many clothes and yet most of us keep doing it. Those are just a couple of mistakes that many of us repeat, and they are usually in activities we don't do very often. We are trained not to make mistakes so when they happen we walk away. Never forget the lessons learned in any stage in any project. They do two things: they create knowledge for the next step and they make us feel better after a bad experience.

GIVE YOURSELF THINKING SPACE

I flew out of Toronto on a November night for a meeting in Rio de Janeiro. The flight schedule required a flight change in Washington, D.C. I boarded the new flight in D.C., due to take off at 9:53 PM. After an hour on the tarmac we realized there was a problem. After another hour we deplaned. At 1:15 AM the airline announced a problem with the GPS and the flight was now delayed to 6:30 PM the following evening.

The good news was we were going to be accommodated overnight at an airport hotel; the bad news was I finally got to bed at 3:00 AM, although it was better than sleeping on a bench in the departure lounge.

Saturday morning arrived and I was presented with a new day in a new environment and space to think. The pressure of modern life makes us view this as a loss of a day instead of seeing this as "thinking space." It is actually a great opportunity. I know we all feel a bit weary when people say "a problem is an opportunity," but when you find yourself stuck somewhere this truly is an opportunity to release your inner thoughts and let them connect and create new ideas.

Some of the world's greatest books have been written in jail! Most of my previous book was written in coffee shops and R&B lounges. Most writers do not leave home without a notebook. Keep a notebook in your car or in your travel bag.

My good friend Nora Camps gave me my first "ideas book." It was bound in Italian leather and was pocket size. Find a book you like that works for you. Nora gave me an interesting tip at the time. She said, "Once you've written down your thoughts, close the book and then at the end of the month read through your 'bright ideas.' "

This may not work for you, it depends on what type of person you are. You may prefer a looser approach. As I write this book I am using a standard note pad; I am in "flow" and a small book wouldn't handle it.

Either way, when you find space, allow your thoughts to wander and transfer your tacit thoughts into explicit thoughts.

THE GIFT OF TIME

Being able to give time to yourself is definitely a skill for some of us. Whether it is taking a bath, watching the sunset, or taking a walk, time and space are something that many of us do not allow ourselves. If you have the gift of being able to give yourself time, then move on to the next page. If not, let me tell you some of what I have learned. When I lived in Europe I did this well. The culture allowed and encouraged "time giving." When I moved to North America, I found that the skill was eroded. The British and the Germans are inveterate walkers. The French and the Italians are café talkers. Either way, it is time out.

The British sport of cricket takes between three and five days to complete one game. Imagine that in North America! If you truly want to discover time and space, go to Africa. An African sunset is one of the world's great experiences.

You have heard the expression "time to heal." Time does that but it also releases those magical inner thoughts and experiences you have accumulated. Steven Covey's famous book *Seven Habits of Highly Effective People* has much to commend, and Covey's personal philosophy is one we aspire to. However, there is a tendency for people to focus on efficiency and effectiveness and forego the 'gift of time' as they read Covey's words *proactive, the end, first, win,* and *focus on 'time efficient.'* My first book was dedicated to my parents and also to my two daughters with the words "to Rachel and Sarah, who remind me to stop and smell the roses." That is a constant reminder to me and we should all do it.

PROBLEM SOLVING

Problem solving is familiar territory for people who have had quality management training but it is traditionally linear in its approach. Nonetheless, people who are left-brain thinkers may want to develop their skills further by pursuing TRIZ techniques, which I mentioned in Chapter 7 "Connecting to the Solution."

Creative thinking is one of the skill sets to be developed by the innovator, and de Bono is of course the master in this field. Lateral thinking is counterintuitive for people who have been through the rigor of Six Sigma training. A great way of breaking the ice here is the Chinese *I Ching.*

Stepping out of the box frequently provides insights to solutions as "time out" changes context. A solution will emerge, buried in the knowledge transfer that occurs during social interactions. The "Aha!" doesn't mean the idea is good. People believe they had a solitary insight, but the real story is that a social encounter was responsible for the idea. The idea that insight emerges unpredictably persists because most people aren't aware of the social encounters that lead to their insights.

Connecting is a key part of problem solving. We are all familiar with the "nine dot" problem or what is often called the riddle. For example, many of us have seen this old riddle, which is a good example of thinking outside the box:

> A man who lived in a small town married 20 different women in the same town. All are still living and he never divorced a single one of them. Yet he broke no law. Can you explain? The answer: the man is a clergyman.

Finding the best solutions then means stepping out of the box and collaborating with others.

STORYTELLING

This has become fashionable in business, and will become vital when a product is launched. In the world of clothing fashions we would refer to a new line of clothing as a "story." The story recognizes that emotion must be uncorked and released for a new product to succeed. There are many books out there on storytelling. It used to be an art. It is now becoming a science. I learned the science when I wrote my novel *The Pointman*.

If you try to convey a conceptual message, most people have to work to integrate the message into the existing body of knowledge in their mind. The mind works like a computer doing a search. It is looking for the right home for the new concept.

Stories, on the other hand, relate to past experience, they focus on the one point that matters, and people instinctively go into connector mode and place the new message into their existing body of knowledge. People are more receptive to stories. They like to be entertained, and stories are more memorable. All of the prophets used parables.

In the world of innovation, wisdom is not conventional. Stories are a great way of communicating the unconventional.

There are distinct stages to a story whether it is a book or an anecdote, whether it is Homer, Shakespeare, or Hemingway:

Act I 1) The Status Quo, 2) The Challenge, 3) The Refusal

Act II 1) The Test, 2) The Crisis

Act III 1) The Resurrection, 2) The Climax, 3) The Return

This is the structure of the classic three-act play, but even the simple anecdote follows the essence of this structure. There are also distinct characters in the story. The Hero, the Mentor, and the Villain actually may embody themselves into more than one character.

If you have the job of developing or delivering a new product or service, I encourage you to develop the art and science of storytelling. When we sold the humble towel at Christy, we told the story of William Miller

Christy, the Sultan of Turkey, and Queen Victoria. People were suddenly buying a piece of history.

OTHER SKILLS TO CONSIDER

Risk analysis is widely understood in business today and is often a subset of project management. However, a basic understanding of risk is useful for the connectors as they evaluate alternative solutions.

Project management skills are definitely required by the developers, and if your business does not have these it is a must to be addressed.

Strategic selling, as opposed to product selling, and development of the value proposition are skill sets the sales people will require. I talked about this in Chapter 10.

The skill of exploration for business partners often resides in the purchasing function and becomes necessary as an organization moves to open market innovation.

Finally, to pull together all of these competencies do not forget that a training plan must be developed that aligns training with the deployment of the innovation plan (see Chapter 19). Remember, adults learn by doing. As you deploy your innovation plan, you integrate the training, experience, and skill development with the deployment plan.

One last point. There is an enormous benefit from being known as an innovative company. You can attract and recruit the very smartest people because they want to hang out with other smart people

As I mentioned earlier in this chapter, I am next going to talk more about networking.

BROWSER'S BRIEFING CHAPTER 13

- Leadership must fully understand innovation in order to implement an effective innovation strategy.
- A one-day workshop will introduce the leadership and also the "I" team to the essentials of innovation.
- The workshop must include practical exercises that relate the learning to the organization of the participants.

Continued

Continued

- The innovation training plan must be synchronized with the innovation plan.
- The early stages of the innovation cycle require the competencies of exploration, observation, interaction, questioning, and listening.
- Note taking is a fundamental skill throughout the innovation process, and technology can help us here.
- Experimentation and willingness to fail are competencies that can be developed.
- Throughout the innovation process we must be able to give ourselves the gift of time and space.
- The skills of problem solving, project management, and strategic selling, which are required in innovation, are generally better known.
- Brainstorming is often taken for granted, yet must be practiced regularly to develop competency.

14
Networking

It isn't what you know; it's who you know.

—Anonymous

I have talked about the technical and process side of networking, but what about the human side?

All through my twenties, in my early years out of university, I remember those parties where we would sit into the early hours of the morning and try to figure out how to change the world. My friends at the time were mostly chemical engineers, like myself, who had joined the research division of a large corporation. I'm not sure whether it was frustration with the corporation or just something we all wanted to do at that age, but all of us said, "Some time I want to run my own business." We were all looking for that niche or "magic opening" that would provide the opportunity. If you have passed the 30-plus barrier you will remember what a traumatic point in life that is. You truly wonder where your life is heading.

I remember that 30-plus point in my life; it was a point in my time with Courtauld's when I was very unhappy with my career path. I had moved from the research division to the textile division and the culture was entirely different. The paradox for me was that I enjoyed textiles far more than chemicals. It was more human and far more artistic and yet at Courtauld's the particular area of textiles I then worked in lacked the trust of my previous environment in chemical engineering.

I made the tough decision to leave the big corporation and start my own business. I decided I could not face myself if I had not at least been through the experience. My parents were horrified. I read a book on starting your own business that included an assessment of whether you were suited to be an entrepreneur. The biggest psychological obstacle to being successful in your own business was if neither of your parents had been self-employed. The odds of succeeding did not look good.

I took a look at what I had going for me. I had learned a lot about textiles and had a good sense of color. I was currently selling fabric to the fashion trade and I was a very good business planner. The other factor was that I had no money.

I decided to set up a business designing and manufacturing women's fashion, an industry that required little capital to enter. I left Courtauld's, cashed in my pension fund, and used half the funds to buy an old car and the other half to buy fabric.

I joked with my friends that I decided what business to go into by asking myself what I liked. I liked clothes and I liked women so I decided to design women's clothes.

NETWORKS OF FRIENDS

If you don't have family support you turn to friends. I had two networks: my original group of chemical engineering friends and a new network of friends at a rugby club I had joined since moving house.

Surprisingly, both networks proved incredibly valuable as I developed my business. You would wonder what a bunch of chemical engineers and rugby players knew about women's fashion. In truth they provided a mountain of opportunities and contacts. The wife of a "friend of a friend" at the rugby club became my design partner. My chemical engineering friends were incredibly curious and kept asking about my new business. Whether you are on your own or in a big corporation your networks are vital.

My networks have changed over time. I no longer play rugby but belong to a dinner club called The Thunderers, to which I was introduced by another friend, Ian Oliver. I no longer live in the UK where my chemical engineering friends are, but we still stay in touch. I have two other major networks: the professional society to which I belong, ASQ, and the global network of ISO, which gives me friends across the world.

Having friends "outside the box" is essential for you to have a fresh perspective on the world. Having an environment outside your own where you can go and experience fresh messages, relax, and allow your mind to unwind is essential.

If you want to join the world of the innovator you need to be mixing in unfamiliar territory. What are your networks?

People in the neighborhood

Parents from your kids' school

A professional organization

A sports club

A hobby club

MySpace and Facebook are not the type of network I am talking about. They are not going to give you that essential human interaction. However, an innovators' Web site such as my www.petermerrill.com can be a great source of innovation knowledge and an opportunity to exchange ideas. I was recently invited to join LinkedIn, which provides excellent points of contact. However, be careful.

It is only when you are experiencing all the channels of communication, written, spoken, body language and signals, that your instincts will work to their full. Only when that happens will you release the subconscious ideas that are deep within you. That is why face-to-face networking is vital.

List the ten people you know best and ask yourself which of them:

- Has a different profession
- Has a different country of origin
- Speaks a different native language

Whatever stage you contribute to in the innovation cycle, you need to be networking. Networking is fast becoming the "new best friend" of everyone in business. The reason? It has reached the magic point in its life where it is on the cusp of moving from art to science. What that means is we are finally starting to understand networking.

SUCCESSFUL PEOPLE WILL NETWORK

In the world of innovation, networking has become fundamental, whether you call it collaboration, open market innovation, or communities of innovation; it is woven through the whole practice and body of knowledge.

Ironically, it was the scientists, or rather the mathematicians, who kept it from us for over 200 years. Euler first developed network theory in 1780, and in Chapter 2 I explained the work of Albert-László Barabási, who brought the science down to earth. At the same time as this mathematical approach flourishes, empirical research is converging with the mathematics to give us a real understanding of how this works.

If you look at the success of some of the great artists in history it was due to their benefactor, who in turn had a powerful network. Amadeus

Mozart, David Hockney, Samuel Pepys: a musician, an artist, and a writer who, although very talented, all succeeded largely through the network of their benefactors and so pushed out their competion. Yes, the ones I mentioned had talent, but there were many others competing with equal talent and of course, "success breeds success."

The other famous story of networking is in Gladwell's *Tipping Point* in which he compares Paul Revere with William Dawes. Paul Revere "knew the right people" and quickly raised a militia, whereas William Dawes, who set off at the same time with the same mission, disappeared into history.

Linus Pauling, the twice Nobel Prize winner, with his famous saying "the best way to have a good idea is to have lots of ideas," openly admits that his success is not due to greater intellect but due to greater networking.

FAMOUS COLLABORATORS

C. S. Lewis first met J. R. R Tolkien at Oxford University. They met regularly and had a common bond in being Christians. However, their perspectives were different. Tolkien argued that the idea of a God made things easier to understand while C. S. Lewis was interested in exploring the "myth" of Christianity. When new ideas developed they would write. The outcomes were for Tolkien *The Hobbit* and *The Lord of the Rings* and for Lewis *The Chronicles of Narnia.*

DIFFERENT TYPES OF NETWORKS

Networks are about sharing knowledge but also about making things happen. If you look at the different stages of the innovation process you see that the emphasis on these two attributes changes.

The first two stages of the innovation process are about sharing knowledge. That doesn't preclude "making things happen" but it does mean a different kind of network. To find opportunities and conceptual solutions, we need a diverse, dispersed network. People who may not yet be customers, people who may not yet be business partners, and definitely people in other disciplines. Remember, the breakthroughs occur at the intersection of different bodies of knowledge. That "spark of ingenuity" (see Figure 14.1).

What we have learned about this type of network is that a large number of people doesn't necessarily mean a good network, and that a sparser, well-distributed network is best in these early stages. The really interesting thing we are learning is that indirect contacts or "two degrees of separation" produce the most valuable knowledge.

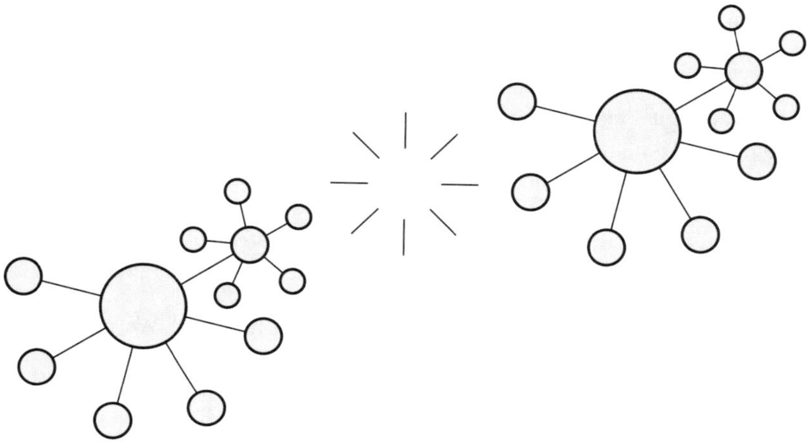

Figure 14.1 The spark of ingenuity.

In practical terms, it has been shown that people who are job hunting are far more likely to find a job through a "friend of a friend" than through any other source, whether direct acquaintances or job ads.

This open and private network has great value when compared to the Internet. It is your private and personal network and so ideas are contained and captured. You can handle intellectual property far more easily when you are the "hub" of this type of private network. On the other hand, if you search the Internet for solutions then it is certain that someone else will have that solution as well. In fact, many others will have that solution.

THE "MIRROR" TRAP

Unfortunately there are many things that militate against us creating this type of open network. People tend to gravitate toward people with similar backgrounds. When this happens people simply reinforce existing beliefs. They just mirror each other's thoughts.

If you have ever worked behind a bar you have seen this. I recall working behind the bar at our rugby club. Throughout the evening I watched people cluster in their familiar groups. As the evening progressed people came to the bar and chatted with me, and I met and talked with more people than anyone else in the room. At a cocktail party the bartender is the hub and usually gains more knowledge than anyone else in the room.

The dense network or "cluster" does have its place in the world of the innovator, but in the later development stage of innovation. That is where the early adopters are engaged and these are going to be "like minded" people. The execution stage will still engage the early adopters plus those who are "potential" users, and this will again be a cluster. The cluster type of network, which is similar to your immediate group of friends, will normally have a very high level of trust and collaboration.[1]

Most personal networks are this cluster type, and so if you are seeking to acquire the new knowledge needed by the creator and connector, your network must change (see Figure 14.2).

We generally lack diversity in our personal network. We repeat this mistake in our organization with our recruitment policy, and engage new people who are similar to ourselves. For successful innovation we need diversity of experience. Let me compare the immediate networks of a couple of people. The first is Hugo, a software engineer in New York, who is from Eastern Europe and enjoys travel. His circle of friends meets regularly at parties, and looks like Figure 14.3.

You can see that only Charlene and Mary have a number of friends outside the circle and Tom is not even well connected inside the circle. You would think New York would provide lots of networking but "turning inward" often happens in the big city. Compare this situation with that of Carol, a housewife in Houston with two young children. You would expect her to have a limited opportunity to network, but she has kept in touch with college friends and met others on her travels (see Figure 14.4).

You can see the diversity of the ethnicity in Carol's group, but also the diversity of activities. She is going to learn a lot from her friends and connect with other people through her friends, who all engage in outgoing activities.

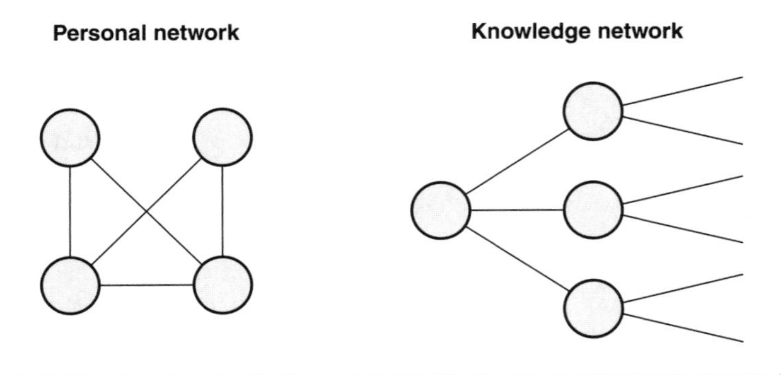

Figure 14.2 Personal network and knowledge network.

	Job	Hobby	Ethnicity	# Friends
Tom	Phone store	TV	NY	3
Charlene	Fashion store	Movies	S. Carolina	10
Dick	Mechanic	Cars	California	8
Pauline	Software	Clothes	UK	6
Harry	Software	TV	UK	6
Mary	Sports store	Skiing	Canada	10

Figure 14.3 Hugo's circle of friends.

	Job	Hobby	Ethnicity	# Friends
Margarite (1)	Housewife	Cycling	Poland	15
Margarite (2)	Mathematician	Chess	New Jersey	10
Renato	Singer	Writing	Argentina	8
Jean	College prof.	Dining out	France	20
Mario	Tour guide	Cooking	Italy	18
Hilary	Tennis coach	Reading	UK	20

Figure 14.4 Carol's circle of friends.

So, how do you develop from the first scenario where everyone knows everyone else to the second, which is diverse and knowledge-sharing? First, let me stress, you don't give up your group of friends! They are important as a key component in your life. However, a "knowledge group" can coexist very happily with them.

YOUR PERSONAL KNOWLEDGE NETWORK

A business network that is diverse and still manageable has 25 to 35 people. Why that number? Well, this is manageable based on experience, and given that I am going to tell you to contact one of the group each business day,

	Job	Hobby	Ethnicity	# Friends

Figure 14.5 Your own circle of friends.

this number also means that the "cycle" is not one month but advances by about ten days each month.

Start by listing your immediate circle (See Figure 14.5).

Now list 35 business contacts that you have. Then go through the exercise of identifying job, hobbies, ethnicity, and, if you have the knowledge, show whether they are a high, medium, or low connector.

Now look at your list and see who the closely matched people are. Decide on who to retain. Your list will probably reduce to 25. Now ask yourself, if you brought your 25 to a party at your house would they mingle happily or would you be concerned about conflict?

If they would mingle happily your group is not yet sufficiently diverse. Not to worry. That will come with time. You're going to be looking for new members in the months ahead.

Let me remind you, networks are about sharing skills and knowledge and sending messages. They are also about overcoming "stove pipes" and barriers to groupings of knowledge and behavior. They are a great way of accessing skills and knowledge you don't have and a great way of sending messages that are important. Your messages may appear to be "public" but when compared to the Internet they are very private.

Let me remind you of one other thing. This is a two-way street, and you have to start the traffic flowing. I recall attending so-called "networking" sessions at the Board of Trade where everyone was "selling" and nobody was "buying." What a waste of time.

You start the traffic by giving. You need to give something interesting. It may be business-related or personal. It will be something you will e-mail and so is likely to be information that a person would value. It may well be different for different people.

Later you may move to an event for your network. I periodically hold a breakfast meeting called the Breakfast of Innovators. It is low-key and there are always a couple of speakers who have an interesting story to tell.

How big can your network grow? Robin Dunbar did some research in 1993 and found an upper limit of 150. The danger above this number is that you lose the "magic touch."

You are starting to see that in the world of the innovator, relationships are a little different. By that same token, leading a group of innovators is a little different. Let's talk about *leading.*

BROWSER'S BRIEFING CHAPTER 14

- Networks must provide human interaction to be effective.
- Networking is moving from being an art to being a science.
- Famous people in history attribute a large measure of their success to their network as well as their personal ability.
- Networks need to be diverse and dispersed for knowledge creation.
- Breakthroughs occur at the intersection of different bodies of knowledge.
- Your private network is more likely to contain knowledge that is exclusive to yourself.
- The Internet contains public knowledge and the knowledge is often not validated.
- Our private networks will tend to include people who "mirror" us, and so will lack diversity.
- To develop networking skill you should cultivate a personal network of 25 diverse people.
- You initiate your network by sending "gifts of value" to the people in your network.

15

Leading Innovation

A leader is best when people barely know he exists,
when his work is done, his aim fulfilled, they will say:
we did it ourselves.

—Lao Tzu

Some people seek leadership, others have leadership thrust upon them. I seem to have fallen into the latter category. My first experience of leadership was when my rugby club, Stockport RUFC, asked me to captain one of its teams. I guess I discovered two of the essentials of leadership: you must have a passion for the activity, which I certainly had for the game of rugby, and you must understand what the team or the organization needs or lacks in order to succeed. In the case of the team I captained, it was simply a matter of ensuring that the team was not short of players. Anyone who has coached or captained a team will know exactly what I mean by that. Ironically, I was far from being the most skilled player on the team but I did have the skill of "reading the game," which is vital when you are in the turmoil of a game of rugby.

THE PERSONAL ROLE

Every leader has a contribution they can make to innovation. Look again at the self-assessment in Chapter 3 for an indication of that role and then where you can actually contribute to the process. My own score is unusual. I am a developer, which comes from my engineering background, but a very close second is my creator score, which is the artist in me. What I have found from empirical experience is that I made my own best contribution by acting as a link between sales and design people. In the Christy business we had people who were much better than me at both these functions. However, as in many organizations the structure did not lend itself to good communication between sales and design. A leader must always be looking for the primary needs of their organization. This way the leader gives the people directly what they want and so strengthens the people's commitment to innovation.

A leader's first task is to understand the innovation process, which I discussed in Chapter 13. Most business writing and popular perception totally misunderstand the process. Business strategists look at the simple model of market development, product development, and product delivery presented in Chapter 1. This model needs more detail in order for the leader to manage it. These additional functions are shown in the revised model in Figure 15.1.

Popular perception goes wrong when it groups activities into sales and marketing, design and development, and operations. The groupings should be 1) marketing with design, 2) development, and then 3) operations with sales.

Every leader has two innovation roles in their organization. One inside the process as a participant and the second outside the process to ensure that it is functioning. I already talked about involving people. A leader must

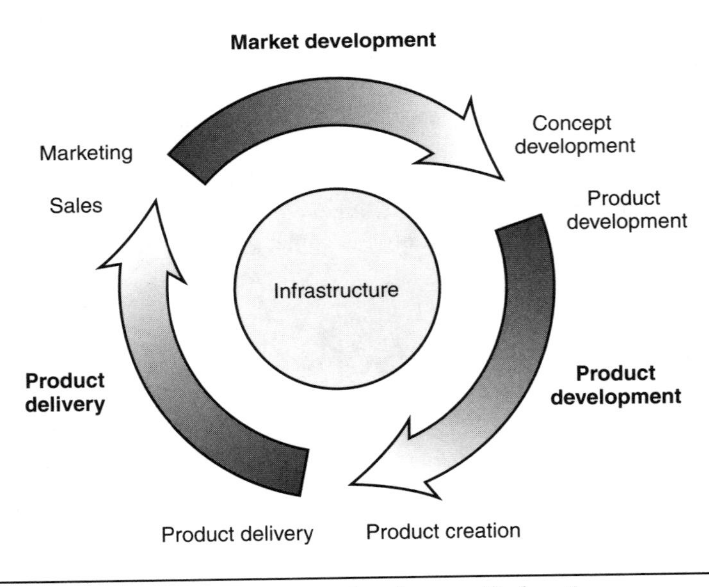

Figure 15.1 Business process areas showing specific functions.

get their feet wet and their hands dirty as they change to an innovative culture. A leader is a change agent. Let me focus on the second role of managing innovation.

SET THE DIRECTION

The second role of the leader comes from our good friend the plan–do–check–act cycle. This gives the leader their mandate. Plan means set direction, involve the people, and give them resources. Strategic direction comes from market trends and opportunities and knowing your organization's core competencies. When I say trends I mean mega trends like technology, environment, and terrorism. These are of course subsets of the biggest trend of all, which is globalization. This broad brief to the market researchers should then leave them "loose" to find opportunities.

The leader must set direction at the marketing stage, knowing the market strengths, such as brand and operational strengths, of your organization as well as your creative strengths, where you should be focusing for your new opportunities.

This direction must not be narrow or you will miss opportunities; Apple would not have gone into the music business if their leadership had thought, "we make computers." On the other hand, the direction must be one that uses your competencies or you will spend forever finding new resources. Apple, again, keeps moving the goal posts so competitors can't catch them.

This is one of the toughest jobs for the leader. Jim Collins in *Good to Great* defined it as the *hedgehog principle*, expanding on the quote from Erasmus.[1] The fox keeps jumping from one opportunity to another. The hedgehog rolls into a ball, focuses, and prevents others from attacking.

RESOURCE THE PEOPLE

Giving time is something most leaders do very badly. The obsession with short-term efficiency is the greatest inhibitor of innovation. People are not going to innovate when they are exhausted in the evening, or at lunchtime when they are hungry.

Innovation is a prime-time activity. Prime time is at 10:00 in the morning, when of course everyone is in routine meetings.

The most overlooked business resource is knowledge. Knowledge comes from education, and education is probably the item that most leaders will cut in a tight budget.

Providing resources means ensuring that you have the right people at each stage of the innovation process. Assess your process (see Chapters 6 through 10) and have your people self-assess. Ask where you are deficient. You may have the right people component but your process may be weak. For example, you may have a strong presence of connectors at the solutioning stage but your process is weak through lack of a defined process or unclear objectives. The leader must ensure a balance of people and process at each stage.

Resourcing also means providing the right tools and technology needed. Be careful here. It is more about developing IT skills and competencies and not about huge investment in the "cool tools" of IT.

MAKE THE DECISIONS

Creating a culture of freedom and choice allows ideas and opportunities to emerge. The tough choice for the leader with lots of ideas is, which ones to run with?

It is so easy to choose based on short-term return. Most businesses look for three years ROI. I mentioned already that Xerox has found that some of their best new products can take seven years to get the ROI they need. Risk must be calculated, then managed. Risk must not be avoided.

The *check* part of the cycle is hard. The leader needs nonfinancial metrics at each stage of the process. A leader with a strong financial background may have the toughest job here. Measuring "exploration" is one of the first challenges. The creators are looking for opportunities. At the outset the metric may be the number of customer interviews, moving to the number of opportunities later. Measuring the connecting or solutioning means, how many potential solutions are you finding?

The selection stage is rich in measurement opportunities. This is where the innovation process connects to the strategic planning process of the business. The leader is now in more familiar territory. Measure failure—remember, Intel said if they are not getting a 10:1 failure to success ratio then they are not taking enough risk. Measure your ratio of long-term to short-term ROI selection. (See Figure 8.1, page 82.) Measure completion time from selection of solution to first sale.

This takes us into the development stage of the innovation cycle. Here the measures are about speed and the more tricky user-friendliness. See Chapter 9. Finally, at the execution stage it is about measuring speed of delivery and measurement of sales growth, again more familiar measurement territory for the leader. I will talk more about measurement in Chapter 17.

TAKING ACTION

The *act* part of the PDCA cycle means the leader must periodically step back and review the health of the innovation process and the organization as a whole. I dislike the word *review*. It is so imprecise. It literally means "look again" and it does not infer taking action. In actual fact, review means compare results with expectations and take action on the shortfalls.

If you were engaging in serious change to an innovative organization I would recommend a quarterly management review of your innovation process. This is where your I-team looks at each stage of the I-process and identifies whether the targets are being met, whether soft targets or hard, whether the process is operating as required, and whether the process is being properly resourced. Your I-plan is then updated. You may choose to do an I-test at points in the process where concerns exist. The typical agenda for the I-review is:

New opportunities

New solutions

Selection of solutions

Measurement of results

Action points in the process

Skill development

Technology needs

PROCESS FOCUS

In the early stages you may be just looking at your I-process and building the process, so the agenda will be structured around process design and resource development.

This is a different kind of leadership and it is at the project level. The job of the leader is to take an I-project from opportunity to execution. They lead from within. Their primary job is to build and sustain the group. Managing the changes in the team's makeup and the team culture is the most demanding task. I will talk more about managing change in Chapter 19, but this is a key issue in an innovation environment. The I-project team will shift in content as you move through the stages of the process. In the first two stages you will be in a loose mode, and monitoring activities is critical. There may be people who have a separate agenda and they may feel strongly about it. They must be allowed to pursue that agenda but not at the expense of the rest of the team. This is where the art of leadership becomes important. Controlling will kill new opportunities.

The leader needs to direct opportunities into the solutioning stage, and there may be solutions that are discarded but which need to be carefully stored for potential future use. The solutions have to be assessed so that useful information goes forward to the selection stage.

The I-team presents its options to the strategic planners, and once selection is made there is radical change. Now we run for the line, and the I-project leader may even change. The team culture changes from loose to tight and the team's makeup shifts to a predominance of developers and doers. This is where the traditional skills of project management kick in.

I was one of the team members that wrote ISO 10006, *Quality management in projects* and I refer you to it as one of the best bodies of knowledge in this area.[2]

DEVELOP THE CULTURE

This is the hardest of all leadership tasks. The looseness required in the creative end of the process and the tightness at the delivery end are in direct conflict, and yet you need both of these modes to exist in the same organization.

Culture is about behaviors, and new behaviors have to be developed. Behaviors always gravitate to the "normal" behaviors of the organization, which are intuitively designed to preserve the status quo. They are designed to avoid wasted time and to increase efficiency. An organization thinking about the future allows itself to think and try out new ideas, and allows itself to fail.

Allowing failure is a tough job for many leaders. Slogans like "Winning isn't everything, it's the only thing" create a fear of failure that in turn creates a self-fulfilling prophecy of failure. A leader must identify the organization's existing strengths and those aspects of the organization that require more work. The aspects that require evaluation are:

- Leadership
- People
- Processes
- Knowledge
- Culture
- Results

This is detailed more fully in the next chapter in the *organizational assessment*. Following the organizational assessment you need to form the core group and educate both this group and the management team in the essentials of innovation. In order to build critical mass you need a short-term win while at the same time developing the vision of innovation. An early win is critical if you are going to make people believe in your desire to innovate. You then start to develop more communities of practice on your road to an innovative culture.

As a leader your immediate actions need to be:

- An organizational assessment
- A management workshop
- Create a core team
- Secure a short-term win

So next let's look at the type of organization the innovative leader needs in order to succeed.

BROWSER'S BRIEFING CHAPTER 15

- A leader needs to know his or her own "innovation aptitude."
- The traditional functions of an organization are grouped with sales and marketing separate from design and development, and both separate from operations.
- The traditional grouping of functions impedes innovation.
- A leader's first task is to set strategic direction for innovation.
- Strategic direction is based on market need and the core competencies of the organization.
- Giving people "creative time" is one of the hardest tasks for a leader.
- A leader must ensure that the right people are involved at each stage of the innovation process.
- A leader makes tough choices on which ideas to pursue.
- Monitoring progress through nonfinancial measures can be a tough change for leaders who have a financial background.

Continued

Continued

- A leader must ensure the switch from loose to tight mode at the tipping point.
- The leader must conduct a regular review of innovation projects.
- Above all, the leader must lead the change to an innovative culture by communicating a vision of innovation and recognizing innovative behavior.

Part IV

The Implementation

16
The Innovative Organization

Once an organization loses its spirit of pioneering . . . its progress stops.

—Thomas J. Watson Sr.

Gerry Kavanaugh makes paddles for dragon boats. Dragon boating is a classic example of a new sport where the people involved in the sport have innovated the equipment they use.

Whether it's mountain bikes, snowboards, or kayaks, the people involved in these sports have seen the need to refine their equipment to make it more effective in its use. This is customer-driven innovation, but you will notice that these are a different kind of customer. They live on the edge. There is a saying, "If you're not living on the edge you're taking up too much space." These people live for innovation because they have passion for their sport.

Gerry Kavanaugh certainly lives on the edge. Gerry is one of my clients and he drove in formula motor racing. While traveling the racing circuit he discovered dragon boating. This sport, which as you would expect comes from Asia, has 22 people in a canoe who race on open water over a distance of 200 to 2000 meters.

Gerry found the sport exciting and got involved. He discovered very quickly that one of the biggest problems in the sport was that given the tremendous force exerted on the paddles, the wooden shaft had a short life, and almost every race experienced "paddle fractures."

Gerry's experience in formula racing had exposed him to carbon fiber, and he connected the two experiences and developed a carbon fiber paddle for dragon boating. His auto racing experience enabled him to develop equipment to manufacture carbon fiber paddles. Gerry's business took off and he quickly had more business than his facility could handle. He won't mind me saying his workshop became hard to manage. Gerry's is a classic startup business.

I have worked with Gerry to help him grow to the next stage in his business. He has developed structure as he grows but retained that flexibility that has made him so successful to date.

A "FLEXIBLE" ORGANIZATION

You may be surprised but Gerry is using the framework of ISO 9001 to get this structure. The standard is giving him the framework to grow his business. He has mapped his business flow and identified his primary risk points and at each of those risk points he is monitoring his processes. He has started a regular performance review of his business and is looking at nonfinancial indicators as well as the financials. He has identified the key skills needed by his people and developed a competence database so that as he takes new people on board they have the right competency to do the job.

Startup organizations are exciting. How many times has the CEO of a large, rapidly growing, and successful business turned to a best buddy who grew with the organization and said, "Remember how it used to be." This is a yearning for those old days when the business was small and hungry, everyone knew what was happening, and you could turn the organization on a dime.

GROWING THE ORGANIZATION

Success breeds growth, and in *Do It Right the Second Time* I explained the limits to growth of an organization. You can generally reckon that for good lateral communication your span is about six people (you may recall de Bono's law). In terms of layers, reckon on three layers or two degrees of separation in the organization as the limit of good communication (see Figure 16.1).

Beyond these limits, your flow of information and hence your flow of knowledge will suffer. What does that mean in terms of the size of a business unit?

Layer one. Six people—that was the core group in the good old days of startup.

Layer two. This was created when each of the original six got a team working for them, again about four to eight people in each team, and you grew to about 40 persons. Over 90 percent of businesses never grow beyond

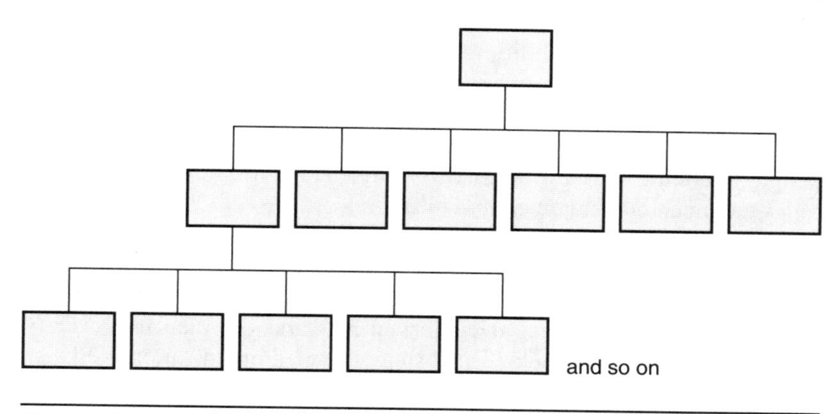

Figure 16.1 As the organization grows, it loses agility.

this. It may be due to a leader who won't let go and keeps the business at this size. Or it may be the absence of business controls, and each time there is growth it lapses back to the "40 limit."

A few go beyond and become an "organization." You will see that the limit here becomes 6 + 36 + 216 or about 250 people. The organization provides a successful product or service and they probably develop a management system that makes things run smoothly.

Therein lies the danger. After three, five, even 10 years of successful operation, the market changes. Some businesses change (painfully) with the market, others do not and they go out of business.

The challenge for the growing business is to identify the new market and product ahead of time, and of course that is what innovation is all about.

THE "AGILE" ORGANIZATION

The traditional organization structure inhibits information flow. For information to flow quickly, we need an organization's processes, its people, and its technology to come together as a system. But this system must be flexible. Innovation is seriously inhibited by a lack of organizational agility.

Throughout the '90s we attempted to address the problem of product flow by viewing activities from a process approach and by process mapping. You will recall from Chapter 2 that process mapping is a technique adopted from the chemical industry. It has a lot of benefits but it treats your organization as if it was a chemical plant and only addresses the product flow aspects of your organization. Nevertheless, if you have process-mapped to identify people's responsibilities then this gives you your process network and also the beginning of your people network.

Let me remind you of Chapters 2 and 14; you will recall that every organization has its own people network, and the information flow between people is often referred to as "networking." However, this network often relies too much on random communication.

The key points in a network are its *hubs* and *nodes*. Examples of these hubs and nodes in the people network of an organization are its managers, its meetings, and social activities, whether formal or informal. In a successful organization formal and informal networks are the same. Albert-László Barabási in his book *Linked* shows how an organization is like an organism; like a human body an innovative organization must develop this network.[1]

Most large companies are better at executing what they do today than changing for the future. A spin-off innovation business rarely cohabits well with its parent. Letting go, borrowing assets, and learning new behaviors are tough tasks.

Many new companies established with a mission of innovation make the mistake of borrowing too much from the parent organization. They borrow too many of the parent's assets and so replicate the old behaviors. The new business does not become a learning organization (see Chapter 2). Innovation is less about technology or creativity than it is about organizational agility, especially at the execution stage. The organization must operate with right-brain creativity in its early stages, but it does need a left-brain structure at later stages to connect it to the main organization because of its need for financial resources.

The secret of success for the startup or spin-off organization is having a network structure, independence, and diverse people.

If your organization is somewhere between the 40- and 250-person mark, you can still develop agility in the innovation process and develop an innovation culture.

Beyond the 250 mark an organization frequently loses its agility and becomes set in its ways. The "elephant can no longer dance." A spin-off organization is often the answer. If you have moved beyond that number there is also a chance that your organization has split like an amoeba into what we appropriately call divisions.

A PAINFUL EXPERIENCE

There is an apocryphal story from Corning Glass that illustrates the pitfalls associated with creating a spin-off.[2] The lessons can equally be applied in organizations of less than 40 people when you are setting up your first innovation team.

Corning saw in the 1990s advances in biotechnology that spawned an industry dedicated to genomics. Researchers wanted to speed up experimentation and Corning saw a significant growth opportunity in provision of glass slides for labs, establishing Corning Microarray Technologies (CMT). They were experienced in supplying to industrial manufacturers, had strong intellectual property rights, and were "world class in glass." However, CMT would be selling to an unfamiliar customer base that emphasized cost and convenience.

The new business shared Corning's development functions and also adopted Corning's rigorous five-stage model for new product development. Management positions were filled by Corning insiders.

As a result, the behaviors and hence the CMT culture were identical to that of Corning. CMT ran into unexpected difficulties, the usual methods for correcting problems did not work, and CMT started missing deadlines.

Two years into CMT's operations they appointed a new general manager from outside the organization who reported directly to the president, and the organization started to follow a looser, more iterative innovation process. Within a year they had a success. The previous GM was highly capable but had been asked to operate in a context where success was unlikely.

NEW KNOWLEDGE, NEW BEHAVIOR

There are some key lessons to be learned from this and other similar experiences.

The spin-off organization has "knowledge generation" and not product delivery as a prime task so it will need to be nonhierarchical with a network structure. This is clearly a given for the creativity that is required. Even as it grows it will need to work on keeping this network. The knowledge it requires will largely come from outside and so it must disconnect from internal knowledge sources. This means a separate IT system. You are creating a new "brain."

You are looking for new behaviors, so this also means disconnecting from the existing HR function.

These are very visible actions and make a clear statement of autonomy. The people who are recruited should be different but they should not all be creative artists. You need to use the assessment in Chapter 3 to create a balance of aptitudes.

One more human resource needs careful thought. The leader. This person is not necessarily a creative genius; they are the person who can draw the creativity out of a genius. Recall my reference to Steve Jobs in Chapter 11. The leader facilitates innovation. They also have the task of getting funding and other resources. The advantage of an outsider is that they do not have the baggage of an insider. However, their disadvantage is that they do not know how to navigate the politics of the existing organization. My experience is that they should report at one level higher, say, direct to the CEO, than might normally be the case. Sadly, one of their tasks will be dealing with tensions, even resentment, if the parent organization has developed a "traditional" culture.

In recruiting an outsider, it will therefore be doubly important to ensure that they "fit" with the core values of the parent. However, one thing to beware of is that the actual behavior seen in the parent organization may not always relate to the original core values of that organization.

Even though the leader has a direct line to the CEO, experience has shown to keep the spin-off hungry for cash. Not so hungry that they are despondent but hungry enough to create a resourceful culture.

MANAGING THE SPIN-OFF

Given that knowledge generation or learning is a prime task, you measure what matters. The key performance indicators (KPIs) need to be nonfinancial and focus on issues like number of market opportunities found, number of potential solutions, speed of resolution of bugs, and speed of client acceptance. The business meeting of the spin-off will focus on these measures and not on profitability. Obviously cost control is important, mainly from the perspective of using it to finance learning.

The business meeting needs to be separate from the parent and probably more frequent. The focus will need to be on trends and not results. This means that plans need to be updated frequently, so predictions need to be made on a regular basis.

THE BRAND

One aspect of the main organization that must be retained in the spin-off is marketing and brand. The marketing people will be delighted, and yet so often they are overlooked. The spin-off typically just engages R&D folks and maybe an occasional production person so they can "keep in touch." Market research should have the key skills for finding the opportunity, and sales and brand people are needed at the delivery end of the process.

What's branding got to do with it?		
Technology	Bell	Intuitive—the phone rings a "bell"
Fashion	Polo	The sport of the rich and famous
Technology	iPod	"I" means "innovation," but also I means me and my personal device
Domestic	Tide	Means clean: clean as the ocean
Leisure	Grey Goose	Ice cold
	Kalashnikov	Violence—brand transfer to vodka
Technology	Quicken	Getting the job done faster
Fashion	Gap	A hole in the wall

Figure 16.2 What's branding got to do with it?

Brand is definitely a key asset that the organization should use if it is the right fit. The Tide brand of Procter & Gamble, although associated with laundry soap, in the mind of the user means "clean." Tide to Go is a cleaning stick. Crest, although initially associated with toothpaste, transferred brand identity easily to toothbrushes.

Think carefully about branding; this is where you name your baby. The name must make a connection for the customer. Making the brand name an ego for the organization is a waste of opportunity. See Figure 16.2.

I have started to talk about the importance of measurement information for the leader. Let's look at this in more detail next.

BROWSER'S BRIEFING CHAPTER 16

- A startup business must have a management system if it is to succeed in the long term.
- In creating the management system, flexibility is essential.
- As organizations grow they become less agile.
- A networked organization is more agile than a traditional organization.

Continued

Continued

- If an organization has grown too large and lost its agility, a spin-off solution may be needed in order to innovate.
- The spin-off should reduce hierarchy, encourage diversity, and be independent.
- The spin-off should develop its own IT and HR to encourage new knowledge and new behavior.
- The leader of the spin-off must reflect the new knowledge and new behavior.
- If the leader is recruited from outside, they may need to report to a higher level to bypass "roadblocks."
- The innovative organization measures nonfinancials and has separate business meetings from the parent.
- The brand of the new product or service must be meaningful to the customer.

17

Measuring Innovation

If you don't keep score, it's only practice.

—Anonymous

I love sport and I love history. Both use measurement in different ways. Sport keeps score and measures through a simple numerical activity, and history measures by recording events. I see people struggle with measurement because they are often unsure of its purpose. People are often asked to record events, rather like a scribe in history, but are not included in the analysis of those events. The scribe often sees events that the stroke of a pen will not capture. Of course, a good scribe takes notes, and I have already emphasized the importance of note taking in the world of the innovator.

Measurement of innovation owes more to the measurement of history than to keeping score in sport. Yes, you measure to improve, and yes, you measure your degree of success. However, you measure initially in order to learn—learn about your process and how well it is performing and what changes you need to make.

There is a cruel twist for the innovator. One of your most important measures will be in order to "learn how well you learn."

So the question is, how well is my innovation working and how do I measure innovation? The common mistake people make is that they try to measure the product whereas they should be measuring the innovation process and system. In Chapters 6 through 10 you assessed your processes, and that was the first step toward a measurement plan.

However, we must take a step back and assess at an organization or system level. In the same way as there are three levels of innovation, there are three levels of measurement: system, process, and product (see Figure 17.1).

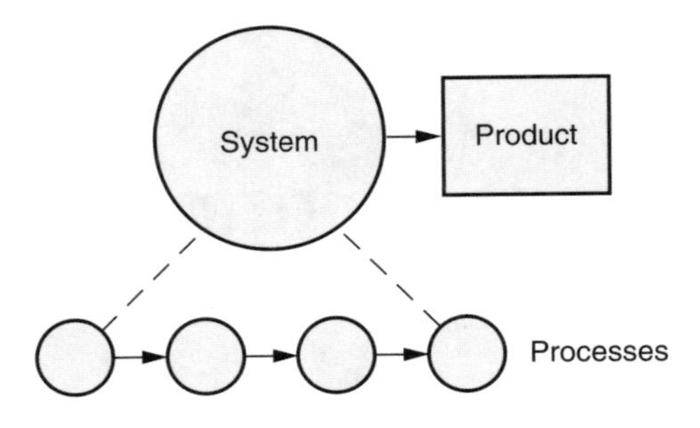

Figure 17.1 Levels of measurement.

MEASURING THE SYSTEM

The ultimate measure of whether your innovation management system is working is whether people buy your product or service. Unfortunately, that is a seriously lagging indicator. We need to measure the system by looking at each of its elements and conducting an organizational assessment.

You need to look at your organization and determine where it is deficient. There are six aspects of your organization you should evaluate. You need to ask questions like the following:

Leadership. Do you encourage risk taking and do you have a commitment to innovation? Does your vision address innovation, and do your leaders and people search for new ideas?

Knowledge. Do you gather competitive intelligence, have knowledge sharing systems, and work with customers and suppliers? Do you have partners in key market areas?

Process. Do you have an innovation process that uses cross-functional teams?

People. Does a high proportion of employees have the capability to innovate, and do they engage in innovation? Is training in innovation offered? Do you encourage skunk works?

Culture. Is innovation recognized and rewarded, and do you document failures to learn? Does your hiring, training, and staff time allow for innovation?

Results. Do you win awards? What percent of sales is from new products, and are products improved continuously? Does innovation achieve business goals?

With the knowledge from the answers to questions such as these you are in a position to start initiating the innovation process while at the same time changing the culture of your organization.

Figure 17.2 is a simplified organizational assessment that will help you conduct this evaluation. You will look at the following elements of your innovation management system: leadership, knowledge, system,

Leadership and Strategy	Strongly agree	Agree	Disagree	Strongly disagree
We make financial investment in innovation				
Our vision and mission includes innovation				
Our leaders take risks				
We target "white space" in the market				
We practice open market innovation				
Total (total each column)	4 ×	3 ×	2 ×	1 ×
Grand total =				

Key issues:

Knowledge management	Strongly agree	Agree	Disagree	Strongly disagree
We collaborate with customers and suppliers				
We share knowledge internally and externally				
We protect intellectual property				
We do not have high dependency on documented knowledge				
We learn from failure				
Total (total each column)	4 ×	3 ×	2 ×	1 ×
Grand total =				

Key issues:

Figure 17.2 Organizational assessment.

Processes and system	Strongly agree	Agree	Disagree	Strongly disagree
We have a defined innovation system and process				
We generate many opportunities and solutions				
We have a selection process for new products				
We make our new products user-friendly				
We are fast to market				
Total (total each column)	4 ×	3 ×	2 ×	1 ×
Grand total =				

Key issues:

People involvement	Strongly agree	Agree	Disagree	Strongly disagree
People know their innovation aptitude				
We develop innovative competencies				
We have communities of innovation				
People are free to explore				
We have a balance of creators, connectors, developers, and doers				
Total (total each column)	4 ×	3 ×	2 ×	1 ×
Grand total =				

Key issues:

Figure 17.2 Organizational assessment. *(Continued)*

people, culture, and results. The assessment asks just five basic questions in each assessment area. This gives a maximum score of 20 points in each section. If you choose, you can multiply each section score by five to give a percent score.

Note that the term "product" is interchangeable with "service."

This assessment gives you a very high-level view of your organization, and naturally each area can be looked at in far more detail. You have

Culture	Strongly agree	Agree	Disagree	Strongly disagree
We are not afraid to fail				
We recruit people who are innovative				
We manage a "loose/tight" culture				
We are an agile organization				
Innovative behavior is recognized and rewarded				
Total (total each column)	4 ×	3 ×	2 ×	1 ×
Grand total =				

Key issues:

Results	Strongly agree	Agree	Disagree	Strongly disagree
We measure learning				
We measure speed to market				
The majority of our products are less than five years old				
We do not need to have recalls or "service fixes"				
We win awards for innovation				
Total (total each column)	4 ×	3 ×	2 ×	1 ×
Grand total =				

Key issues:

Figure 17.2 Organizational assessment. *(Continued)*

already been to the next level of detail for *process* when you did your process assessments in the earlier chapters. However, you should at this stage list the key issues from each section of the assessment. This tells you where work is required. See Figure 17.3.

Your assessment so far is quite subjective. You have viewed the primary aspects of your organization and allocated an arbitrary score. Nevertheless, it is a form of measurement. Measurement is really an attempt to present history in the language of the scientist.

• Leadership and strategy
• Knowledge management
• Processes and system
• People involvement
• Culture
• Results

Figure 17.3 Key issues.

An ISO 9001 gap analysis or a Baldrige assessment looks at an organization from a system perspective, and its real value is to present an organization with a series of required actions that can then be developed into an action plan. A more detailed version of the assessment on the previous pages will enable you to do this. Let's move to the next level of measurement.

MEASURING PROCESS

Naturally, the essential rules of measurement apply. First, identify where the risk in your processes is highest. You will find that from the assessment tool in the earlier chapters. Good examples of measurement opportunities might be the ability of the initial opportunity stage to generate ideas or the ability of the final execution stage to get to market fast. People usually go wrong by measuring the outcome or the product or the number of new products or patents produced. Other measurement examples might be the number of breakthrough ideas we develop, the performance of our relationships with inventors, partners, or suppliers, and the impact of our innovation efforts on our customers. These measures help us understand and improve the processes that support innovation.

As companies grow, they use measures as a way to evaluate performance. What starts as a few key measures leads to a "mountain of measurements." This mountain can lead to a focus on "completing the list" instead of using measurement for improvement. Measures are fun to create and difficult to destroy. Review measures quarterly to see if they are still relevant. If they're not, eliminate them.

Earlier in this chapter you assessed where your company is on the innovation totem pole. Measure where you are weak. Drill down with your measures. You measure to learn about something you need to better understand and then to improve. Where your high-level measures don't tell you enough, go deeper. An initial measurement plan for your innovation process might look like Figure 17.4.

Process	Owner	Primary objective (example)	Performance measures (example)
Opportunity	Marketing	Find new opportunities	Percent of new opportunities in new market
		Find market "white space"	
Solution	Design	"Out of box" solutions	Number of radical solutions
Selection	Leadership	Narrow focus	Number selected
		Take risk	Number of failures
Development	Development	Make user-friendly	Behavior change
		Rapid completion	Time to complete
Execution	Production	Speed of delivery	Speed
	Sales	Customer acceptance	Customer uptake

Figure 17.4 Innovation process measurement plan.

MEASURING PRODUCT AND SERVICE

Measuring the later stages of your innovation process is not too difficult. You may well be doing this already. Software developers and IT technicians are on very familiar ground when measuring things like "first time fix" and "errors per line of code." The same concepts apply at the development stage of any new product. At the execution stage, sales and production people are familiar with measuring speed to market, speed of acceptance, production errors, and service errors. My chapter on measurement in *Do It Right the Second Time* describes how to measure at these final stages of the innovation process.

It is in the early stages of the innovation process where the measurement challenges arise.

You will notice that I showed fewer measures in stages one and two of the process. This is where your measures are more difficult to establish and need to focus on process inputs. At the opportunity stage you need to monitor trending of the mega trends. Your market research people should do this. You need to measure both customer and market opportunities identified, and also the impact of those opportunities in terms of time and money wasted.

At the connector stage you need to be measuring the number of potential solutions for each opportunity, the behavior change required for a given solution, and the risk attached to each solution. This is partner risk and delivery chain risk. These are issues I described in Chapters 8, 9, and 10.

THE MEASUREMENT PLAN

To be sure you have covered all the bases, review the model in Figure 17.5 and see if there are any areas of your management system where you are not monitoring.

Monitoring means just "feeling the pulse." Measurement means collecting data and analyzing it to gain knowledge. In Chapter 4, I explained that:

Data are just numbers.

Information is patterns in the data.

Knowledge is information that can be acted on.

Remember, you are measuring to see if your innovation efforts are bearing fruit. If they are not, you need to act. Measurement is about improving your ability to innovate.

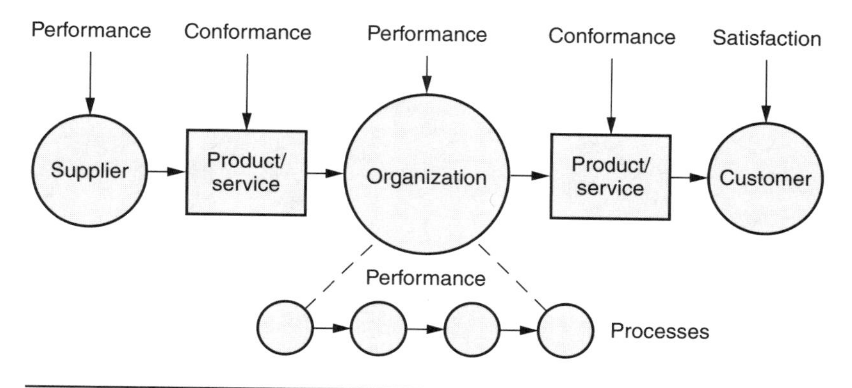

Figure 17.5 Areas for measurement.

As the innovative organization develops its ability to innovate, it must then expand its horizons to *open market innovation*.

BROWSER'S BRIEFING CHAPTER 17

- You measure innovation to see if it is working.
- The common mistake is to measure the product of the innovation process.
- Measurement should focus on process and system.
- The areas to evaluate in your system are leadership and strategy, knowledge management, processes and system, people involvement, culture, and results.
- You should at some point conduct this assessment at a detailed level in your organization.
- The organizational assessment enables you to focus on areas of weakness and build an action plan.
- Process measurement should occur where the innovation process risk is highest.
- Focus on process inputs to avoid lagging indicators.
- Develop a measurement plan.
- Review the list of measurements on a regular basis and cull those that no longer have value.

18
Open Market Innovation

Knowledge is of two kinds. We know a subject ourselves, or we know where we can find information upon it.

—Samuel Johnson

My university thesis was on "The Plasma Jet As a Chemical Reactor." To this day I recall my primary finding was that progress in the research and development of this concept had been painfully slow because of a lack of collaboration. The progress of the plasma jet, creating the fourth state of matter, never got off the ground because it was driven by a "trading" philosophy and not a research philosophy. In business we have an instinct to trade, in research our instinct is to share. I joined Courtauld's research division after a school and university life of sharing knowledge. It seemed only natural to continue.

As a chemical engineer I went into an industry where vertical integration and complete secrecy were the prevailing principles. Industrial espionage and patent protection were the norm. One of my own company's new product developments was an acrylic fiber. A sodium thiocyanate solution was the solvent for the acrylic and it wasn't a great leap of science to move from the potassium-based salt of the competitor to the sodium thiocyanate that my own company used. R&D labs worked this way. You could say it was innovation, but it certainly wasn't "breakthrough" innovation.

There is more knowledge "outside the box." There are limits to the innovation that can be achieved by working with today's customer. Once your innovation process is working, you need to extend it. Radical innovation comes easier with knowledge from outside. Products and services are developed faster and cheaper. If we want breakthroughs, the chance is that we will not have the necessary body of knowledge within the boundaries of our own group of people or even our own organization. We also know that the major innovation breakthroughs occur at the intersection of disciplines. However, people know that knowledge is power and so people retain knowledge and don't share it. On the other hand, in our world of astronomical knowledge growth it is no longer possible for one person to possess the knowledge that an organization needs.

One of the most convincing pieces of evidence for open market innovation came from IBM in their global CEO survey. The chart in Figure 18.1 shows how CEO respondents selected the most significant sources of innovative ideas.

The number alongside each category shows the percent of respondents selecting that particular source of ideas as "primary." Each respondent was able to make up to three choices and therefore a maximum theoretical score would be 300 percent. Not everyone made three choices.

This survey shows overwhelming evidence that collaboration between employees, business partners, and customers produced the best innovation results. That is what open market innovation is all about.

The IBM global survey of CEOs showed a devastating indictment of R&D labs and a remarkable shift to external sources of ideas and expertise. We need to recognize that the smartest people don't work for our own company, and with today's search abilities, finding smart people is much easier. The social interaction with others and the interaction with other disciplines are what generate both opportunities and solutions. You do not need to own the expertise.

This is what has led organizations such as Apple, Procter & Gamble, and IBM to develop their open market innovation. The premise of there being more knowledge outside the box is not new. In times of rapid innovation, knowledge lifetime shrinks. As far back as the Renaissance, which was a time of rapid innovation, the textile industry in Tuscany recognized that if you offer knowledge to others you get new knowledge in return. The textile industry has worked this way for centuries and still does. Li & Fung,

Employees 41%	Consultants 22%	Internal sales 17%
Business partners 37%	Competitors 20%	R&D 16%
Customers 35%	Associations 18%	Academia 14%
113%	**60%**	**47%**

Figure 18.1 Sources of innovative ideas.

the Hong Kong clothing manufacturer, has over 7000 other organizations in its network.

HOW OPEN MARKET CAN WORK

Networks of innovation create far more value than the closed model of innovation.[1] Open market innovation has been successful in areas such as open source software development, but companies are unsure of the best way of operating with the open model of innovation.

One method is to use innovation go-betweens or headhunters to facilitate the exchange of sensitive information between companies. They connect interested companies with sources who can provide solutions to innovation challenges. You hire them to share information about your innovation needs with agents representing other companies and to help structure your engagement terms.

People talk about open market innovation and how if you step outside the box you will find lots of new ideas. And yet little is said about how this open market works. Much is said about "networking externally," but with whom? Is it your customers, your suppliers, your competitors, or totally different industries?

If you look at the stages in the innovation process it becomes clearer who these partners should be. Remember, breakthroughs come from the intersection of disciplines.

Picking your partners is clearly a critical activity.

Stage 1: Creating the Opportunity

Sales people in your own organization mix with sales people in other industries. If they don't, they should. Your market research people should be looking at industry trends but also mixing with customers and consumers. Remember, the mission here is to find an unsatisfied need. Remember also that the best opportunities come from secondary connections. This network needs to be dispersed.

Stage 2: Finding the Solution

This is where cross-connecting with other industries becomes essential. Remember, Henry Ford's idea for mass production came from a meat processing factory.

The link between processing lots of cars and processing lots of meat is in the word *processing*.

The skill of exploration becomes more demanding here. The opportunity guys were explorers but the solution guys have fewer choices.[2]

Stage 3: Making It User-Friendly

This is where customer and consumer networking becomes critical. There are people downstream who will "say it as it is." You need those people. You are testing solutions to find what works. You don't have a lot of time. Knowing the right people to talk to means you must have been listening to what the stage two people were doing and prepared your network. These are your early-adopter customers that you are working with. You actually knew who they were back in stage one when the need was identified. This is also where you develop your supplier network.

Stage 4: Execution

Here your network becomes about degrees of separation. One degree doesn't do it. The customer's customer, at the very least, and three degrees is preferred. What is holding up early adoption? One degree of separation will mask the truth.

THE REWARDS

As one example, Procter & Gamble (P&G) set a goal for 50 percent of innovation to come from outside its organization. They have been very open about the results. R&D productivity has seen a 60 percent increase, their innovation success has doubled, and R&D cost as a percent of sales has dropped from 4.8 percent to 3.4 percent. They have produced 100 new products in two years and their share price has doubled.[3]

Apple took the iPod to market but PortalPlayer created the network or "iNet" that enabled the product to be developed and manufactured. They used vendors from different "ecosystems": disk drive from Singapore, semiconductor from Taiwan, and software from Bangalore.

There are many exciting stories of external collaboration. However, be careful. To quote one CEO, "having a few beers together is not collaboration." Collaboration is a discipline. The IBM CEO survey showed that external collaborators outperformed the competition in both revenue growth and margin. Another interesting statistic in that survey was that public sector organizations collaborated externally to twice the extent of private sector organizations. James Moore in his 1996 book *The Death of Competition* talked of the importance of creating a business "ecosystem."

Since then we have seen the concept of alliances evolve dramatically. However, we have at the same time seen that the majority of alliances are doomed to fail. How do we overcome that problem?

MAKING ALLIANCES WORK

A major reason for the failure of alliances is the focus on the legal and financial aspects of the alliance to the total exclusion of the relationship and resourcing aspects. Building trust is essential, and that trust builds when behavior follows expected patterns and patterns that the other party understands.

As a result, the method of working needs careful thought. Some people are comfortable with Skype, others prefer e-mail. Some are happy with conference calls but others prefer regular face-to-face meetings.

It is far too easy to lay out generic principles in the early stages of relationship-building. The principles must be specific. Avoid words like *review, appropriate, frequent,* or *necessary.* The decision-making process in each company needs to be understood by the respective partners. At what level are decisions made and what are the steps in the process? What are the key decisions that are likely to be taken as the project moves forward?

In the same way that metrics for an innovative spin-off should be nonfinancial (see Chapter 16) there need to be nonfinancial leading indicators for the open market alliance. Obsession with short-term financials will kill the relationship.

Using the type of process measures described in Chapter 17, such as number of opportunities, number of solutions, and speed of bug elimination, will drive progress.

SEEK TO UNDERSTAND

An alliance is normally formed because each party brings an asset, whether it is market, intellectual property, or technology. Recognize that the other partner will not understand your technology if that is what you bring to the party. If the other partner brings market opportunity you need to take time to understand that.

This is only the first step. Differences in culture are the big issue. What might at first be seen as "attention to detail" can rapidly become "slow and bureaucratic." A behavior that at first is seen as "fast response" can soon be seen as "reckless."

In my own work in ISO, I work with cultures from across the planet. Europeans seem to want to debate forever, while North Americans race on with thoughtless abandon. East Asians are inscrutable, and people wonder what they are thinking. Taking time to form working relationships in our meetings is a critical part of our work.

Fundamental to relationships is agreeing how information will be shared and how problems are shared. There need to be agreed approaches for both of these as well as for normal work activity. Remember also that problems may well arise from within your own company, and you have a responsibility to address them internally when that happens.

NETWORK ALLIANCES

The next level of behavior in open market innovation is *network innovation*. This is where your alliance is no longer with two, three, or four other businesses, but can be with a myriad of other organizations. This is like forming a community of innovation, which was described in Chapter 12, but where the members of the community are all outside your organization. The issues of trust and relationship-building now multiply exponentially.

You need a network coordinator who will recruit participants and talents just like in the community of innovation but on a larger scale. You need a "context for participation" such as chat rooms supported by Skype. There is a loose mode of operation in the early stages, and trust is key, but there are tight action points just as with internal innovation.

The coordinator should not be alone but should have a small core team of three to four persons. The exact same issues are addressed as in forming alliances, but you probably need to keep agreements simple as many participants in the network may well be "lone wolves." Clear timelines become important but should be balanced with freedom for participants between the "action points." Interaction between two or more members of the network is a clear challenge for the coordinator, and showing participants where they fit in the network is another part of building trust.

REAL-LIFE RESULTS

One of the more "fun" stories at P&G is of the "printed Pringle." We all know Pringles. They are potato chips for neat freaks. They certainly lack excitement. The innovative breakthrough was transferring the idea of the fortune cookie onto the Pringle. Fortune cookies would be boring like Pringles if they didn't have a message in each cookie. The old P&G would have spent five years developing the technology for printing vegetable dyes. The new P&G bought it. They went on a global search and found a baker in Italy who had developed the technology for printing vegetable dyes onto bread. The baker loved the money and P&G got to market fast. As they say in P&G, "We have replaced 'Not invented here' with 'Proudly found elsewhere.'"

IBM Alphaworks looks at open market innovation from a different perspective. Contrary to the image implied in Apple advertisements, IBM (a PC manufacturer, among other things) is a highly innovative company.

You only have to look at their record going back to the invention of the functional ATM, their work on artificial intelligence, and the number of patents and scientific prizes the company has generated. So productive is their work in the field of software alone that they were feeling the frustration of their innovators in the 1990s as their ideas were failing to see the light of day.

The company set up the Alphaworks Web site, which offered free 90-day trials on unused software. The response was phenomenal as developers and businesses downloaded the software. But here is the twist: ideas came to IBM from their potential customers. Over 40 percent of the products going on the Web site are purchased, and IBM now earns $2 billion each year from royalties on its patents.

In the last three chapters I have introduced you to the implementation of innovation: the type of organization you need, how you measure progress, and the longer-term move to open market innovation. There is one other secret ingredient to implementation. Getting an early win and lighting the fire. I will explain this in the next chapter.

BROWSER'S BRIEFING CHAPTER 18

- Radical innovation is achieved more easily with ideas from outside your organization.
- Breakthroughs occur at the intersection of disciplines.
- Knowledge lifetime shrinks in periods of rapid innovation.
- External networks will change at different stages in the innovation process.
- Networks can be with customers, other industries, suppliers, or consumers.

Continued

Continued

- Collaboration must be disciplined.
- Alliances succeed when attention is given to relationships and resourcing, not just finances.
- Take time to understand how an alliance partner shares information and solves their problems.
- Network alliances are less formal.
- A network coordinator is a key player.
- The coordinator recruits participants, builds trust, creates the context for participation, and ensures tight action points.

19
Lighting the Fire

The time to hesitate is through
No time to wallow in the mire . . .
Come on baby, light my fire.

—Robby Krieger and Jim Morrison/
The Doors, "Light My Fire"

I recall a company in England called Ashton Brothers. They made a product called Zorbit nappies, or what in North America would be called diapers. They had the most powerful brand in the baby care industry. They had been incredibly profitable, year in year out, for nearly a century. The CEO was an accountant and rejected the innovative move to disposables. This could not be justified given their huge investment in terry weaving machines. That investment had to be recouped. Terry cloth was "what we do well" at Ashton Brothers.

My own company, part of the same corporation, looked on aghast. The global trend to disposables was moving like a juggernaut.

The Zorbit nappy business, after nearly a century of success, crashed and burned in two years. The brand disappeared overnight and became synonymous with "yesterday." The new generation of mothers no longer was interested in hanging a dozen bright white terry napkins on the washing line in the back yard. Washing nappies was no longer an "act of love" for the newborn. It was now domestic imprisonment.

You have to determine what is the social change affecting your own market. You have to talk to your customer and use those questions from Chapter 6 to find out what is really bugging them. The evidence may eventually show in the declining revenue of one of your products, but by then it may well be too late. It was too late for Zorbit.

THE I-TEAM

The I-team members are the change agents. The job of the team is twofold. First, to initiate the innovation process, but at the same time to drive the culture change that the organization must undergo. You will have a good sense of the changes needed from the organizational assessment in Chapter 17 on measurement.

You will recall from your own self-assessment earlier in the book that you need representation of each of the innovation aptitudes at each stage in the innovation process. By that same principle the I-team also needs a balance of creators, connectors, developers, and doers.

The team should represent all functions from marketing through R&D to production and sales. The people on the team should represent all levels in the organization, but wherever the team member is not a functional head, the team member has to command the respect of their functional head. In an organization going through "innovate or die" or in survival mode the team will need a high proportion of the senior management team.

Later, some members of this team will probably also become members of skunk works or communities of innovation. The team leader is either the company president or VP–marketing.

One of the early activities the team might conduct is an "I" day, or rapid innovation, to initiate the whole process. However, I will talk about that later. The biggest challenge the team has is to overcome resistance to change.

CHANGING TO AN INNOVATIVE CULTURE

Resistance to change to an innovative culture is especially common if the organization has a history of success. It will be necessary to create a sense of need to change, and the strategy for handling change should follow the approach I outlined in Chapter 11.

The *road map for change* (see Figure 19.1) develops John Kotter's eight stages for leading change using a pictorial approach. Change is not a linear process, but a picture enables us to capture a sense of sequence. You might for example, create the I-team before creating the sense of urgency. You may well raise the bar before creating critical mass. You will certainly be resourcing the process throughout, although this becomes critical when the mass is critical. You will notice I have split the model into *people* issues and *process* issues in line with the treatment in this book, but again that is

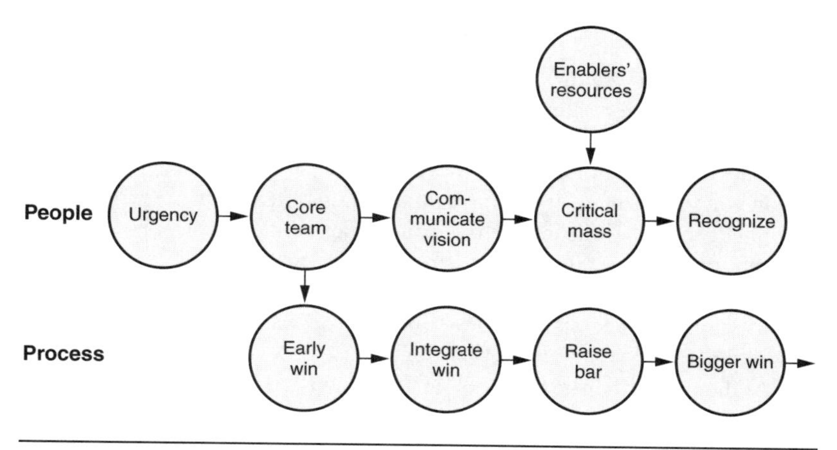

Figure 19.1 The road map to an innovative culture.

not absolute. Use this map as a simple visual to determine whether you are addressing all the key issues in the change process.

Let me take you through the stages in the map.

CREATING A SENSE OF URGENCY

Identify falling revenue or market share with a key product or service and develop a sufficient desire for action. Pick a high-profile item or customer. People must see a need to innovate.

Unless you seriously frighten people they will get on with "business as usual." Kotter calls this "creating the burning platform." You must make people believe that "unless I take action I will die." This is one of the toughest of all tasks in the change process. People inside the business can be happily filling in forms and churning out widgets while flames are engulfing the building. People outside the business just cannot believe the people inside can be so stupid.

THE JOB OF THE I-TEAM

The team will include people from outside management and must be designed carefully.

The I-team works to a project plan (see Figure 19.2) and typically meets monthly to identify resource issues and progress issues. They also use the road map for change to monitor progress.

Rather like in the innovation process itself, the I-team will use non-financial measures. Clearly you need a team of "believers," and for that reason you may wait until you have done the organizational assessment and the management workshop before you make your choices. These people must not only believe, they must carry the respect of others in the organization. However, there is a risk in selecting all doers. You need that mix with creators, connectors, and developers if you want a real perspective on innovation.

At the outset the I-team picks the early win and gets the support of the senior management team. The team manages communications as that early win moves forward, and identifies innovation allies in the business. These allies are vital in the effort to build critical mass. The team should use project management techniques in the first 12 to 18 months to plan, monitor, and control progress.

One other tough task, aside from "lighting the burning platform," is the need to monitor the project team's shift from loose mode in the earlier stages of a project to tight mode in the later stages.

EARLY WIN

An early win is critical if you are going to make people believe in your desire to innovate. "Low-hanging fruit," items most easily attainable with high benefit, are identified. These wins must be "created" and not just based on hope. The win must produce a quick result and then be publicized. You use a rapid version of the innovation process to achieve this. The early win will not be a modification of a declining product; it should be a new product with a customer with whom the organization is experiencing declining revenue.

The "I" day is a carefully orchestrated day designed to produce rapid decisions and give high visibility to the innovation process. It is a day in which you should include a key customer or maybe two from the area in which your opportunity exists. If you have identified a product that is dying or a service with serious problems, they will be delighted to be involved. Naturally, you will avoid a customer from whom competitors could learn about your idea.

A word of warning about this early win: keep a log of events, decisions, and actions. This is often overlooked in the "first fast" project. This will be an important learning experience. You will need to do a lessons learned

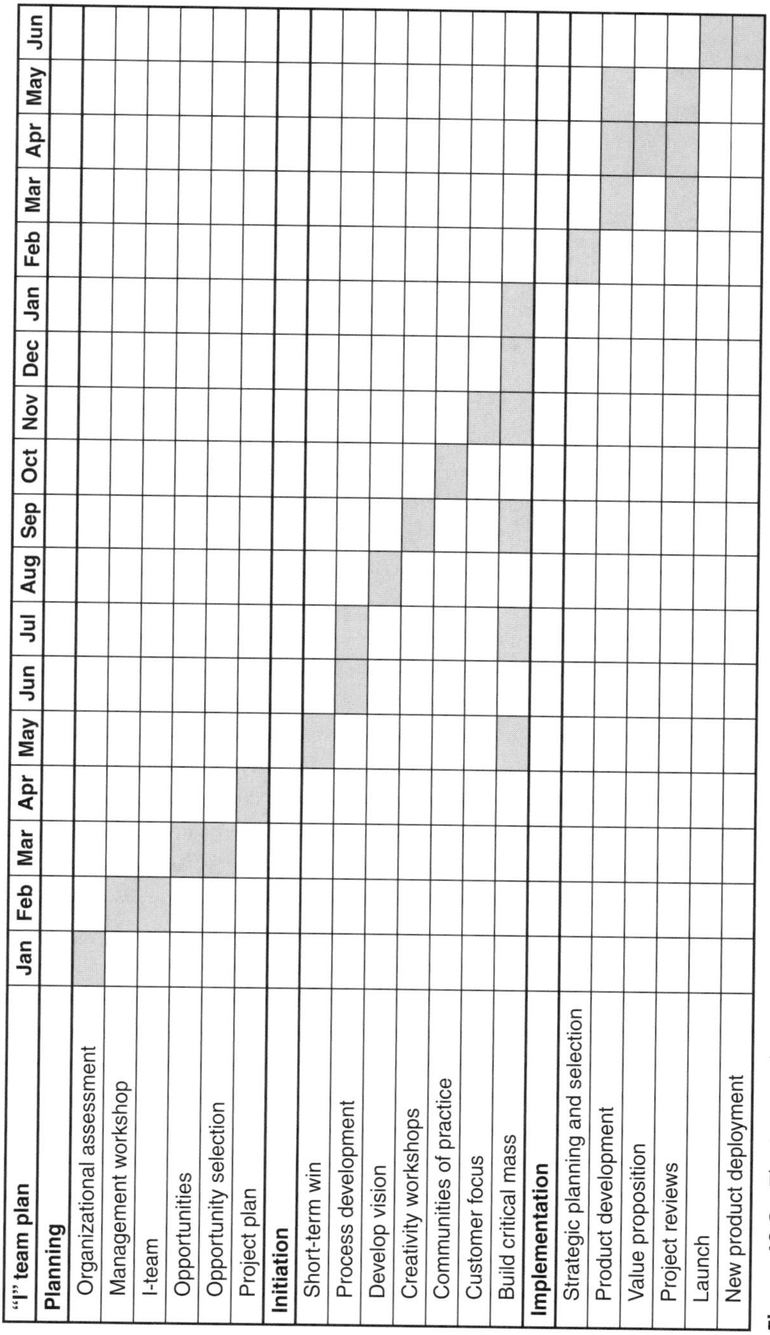

"I" team plan	Jan	Feb	Mar	Apr	May	Jun	Jul	Aug	Sep	Oct	Nov	Dec	Jan	Feb	Mar	Apr	May	Jun
Planning																		
Organizational assessment																		
Management workshop																		
I-team																		
Opportunities																		
Opportunity selection																		
Project plan																		
Initiation																		
Short-term win																		
Process development																		
Develop vision																		
Creativity workshops																		
Communities of practice																		
Customer focus																		
Build critical mass																		
Implementation																		
Strategic planning and selection																		
Product development																		
Value proposition																		
Project reviews																		
Launch																		
New product deployment																		

Figure 19.2 The I-team plan.

session once the early win is out in the marketplace. I will describe the "I" day more fully later in this chapter.

COMMUNICATE VISION

The I-team must draw in the people by focusing on the two or three most important successes. These successes must be easily communicated in concise and simple wording. They must keep repeating the successes.

A vital tool in doing this is the "one minute message." Phil Crosby used to call this the "elevator speech." The term comes from when Phil was VP of quality at ITT. One day he entered the elevator at the ITT building along with Harold Geneen, the CEO of ITT. Harold asked Phil the simple question, "How's quality, Phil?" In the sixty seconds it took the elevator to reach the top floor of the ITT building, Phil Crosby told Geneen what was wrong with quality at ITT, how to solve it, how much money could be made, and importantly, what he wanted Geneen to do. The "motivational sequence" flows like this:

1. Gain attention
2. Visualize need
3. Answer need
4. Visualize result
5. Call to action

The elevator speech in your own company will be entirely different from the speech in other companies. If your mission is to recruit people to join communities of innovation it could look like this:

> We have been making *Z* for nearly a hundred years, but we all know from the global data that the product will die in the next two years. We need to replace this product with one that meets the needs of tomorrow's family and not today's. Our innovation day identified three critical projects that need immediate execution to achieve that change. If we execute these projects successfully we will open new business with 10 major new customers over the next year. I would like you to join the project team that is working on project Z+.

Less than 100 words is your target. Every member of the I-team must have the team's elevator speech in their pocket.

Building innovation into the company's vision is a task led by the president.

BUILD CRITICAL MASS

Communicating vision is one of the toughest tasks. Seventy-five percent of the people must become believers in the first year, a 3:1 ratio. Storytelling, which I described in Chapter 13, is one of the key skills here. Stephen Covey in his book *Seven Habits of Highly Effective People* describes the second habit, "Begin with the end in mind." He describes the building and sharing of a vision.

The strengthening of the stages in the innovation process is the job of the senior person or persons in that functional area: at the *creating* stage the VP–marketing, *connecting* stage the VP–quality, *developing* stage the VP–R&D, and the *doing* stage will have both operations and sales working in tandem. They are key players in building critical mass.

After the early win, the lessons learned need to be applied to future projects. The organizational assessment should be revisited with a fresh perspective. Now is also the time to clearly define your innovation process. This is in anticipation of raising the bar and ensuring that in the future people follow the process.

One of the issues I talked about earlier in this book is how the senior people acquire mind-sets over the years, and even though they accept intellectually the need to eliminate old products, you're probably talking about something that was once "their baby." Passive aggression, enemies "laying in the weeds," and silent resistance are the greatest enemies of change. These must be addressed and brought out in the open. This is the antithesis of vision. It is finding out what people want to hang on to and addressing the issue.

ENABLE ACTION

People must be given the authority and time to innovate. Leaders must find the obstacles and remove them. Compensation should be structured to support innovation. Remember, the secret is to reduce hierarchy, increase autonomy, and encourage diversity.

The popular view of innovation is that it is just "technical stuff." It is actually far more "people stuff." The leadership role now starts to entail not just commitment but also involvement. The senior leadership will need to deal with politics and kill any attempts to stop change. They will need to ensure that skunk works activities are properly resourced. Resourcing means ensuring that people truly have time to do the project work. At this stage we are starting to see ownership of innovation move away from the I-team and into the hands of senior leaders as we build critical mass.

RAISE THE BAR

The short-term win creates danger as people relax. There are more ideas outside the box. Open market innovation is your next destination. When you "raise the bar," you enter into more ambitious projects as well as engaging in open market innovation with your business partners.

Continual culture change is essential until innovation becomes business as usual. Culturally, the "next generation" must personify innovation. New people entering the organization must possess the values and behaviors of innovation described in Chapter 11. The innovations you produce must be explained to both these people and people already in the organization.

INTEGRATE THE WIN AND THE CHANGE

Linkage of the innovation process to the business strategy becomes vital at this stage. The *selection* stage of the process is owned by the leadership and strategic planners. The business now needs to look outside for strategic partners as the bar is raised.

For information to flow we need processes, people, and technology to come together as a system. But this system must be flexible. Innovation is seriously inhibited by a lack of organizational agility.

The organizational assessment is repeated in full and the defined innovation processes are also reassessed for weaknesses.

RECOGNIZE SUCCESS

I described this in Chapter 11. The behaviors to be endorsed are identified and new behaviors are recognized, not heroic defenses of the old. Building trust reinforces new behaviors when they are based on the new values. Recognition and reward must be aligned.

Throughout the stages I have described, a comprehensive recognition system needs to have been operating.

An early win is critical if you are going to get started and are going to make people believe in your desire to innovate. An excellent way to do this is through an "Innovation Day." The "I" day is a condensed version of the tipping point or strategic planning process.

THE "I" DAY

The "I" day must be facilitated and involve both the management team and the people. This rapid innovation is initiated by the I-team, which selects potential areas of opportunity and charts the issues. This is similar to stage one of the innovation process. The team plans the Innovation Day agenda and shares team tasks for that day. Importantly, they also prepare the follow-up to that day.

There must be a senior sponsor for the "I" day, ideally the company president. The sponsor will make the final decisions on "I" day and also hold people responsible for actions after that day. There will also have to be members of the senior management on the decision team and they will need to be selected. The sponsor will need to be prepped by the facilitator on the decision-making process and criteria.

Brainstorming

The "I" day starts with people brainstorming opportunities using "the wisdom of crowds" and identifying the best opportunities. You will be amazed at how much hidden knowledge and experience can be unearthed in this process.

There will be three or four brainstorming teams. The teams create "idea charts," which they then present to the other teams. The groups then vote on the three or four best ideas from each team, and duplications and overlaps are resolved.

This is followed by brainstorming of potential solutions for the opportunities identified.

In addition to potential solutions, they develop plans for implementing the solutions. The solutions should be quantified and a cost/benefit analysis and a risk analysis performed. Out of this you can agree on the best innovation solutions to move forward. The outcome of the exercise is a priority list of the primary opportunities in your organization. These opportunities are "owned" by the people and there is desire for action.

The Town Hall Selection

Solutions are listed on a flip chart, showing resources requirements, time frame, ROI, and risk. A standard chart format is important to facilitate the decision-making process. These solutions are then presented at a town hall meeting. Each team presents its opportunity and solution together with ROI and risk. These may not be radical innovations at this early stage in the innovation journey, but there will likely be some "sacred sacrifices," and this will be a test of the leadership.

Rapid decisions are then made by the senior manager based on the cost/benefit analysis of the solution and the feasibility of the plan.

There is a Q&A session and then the sponsor makes a decision on the spot. The reasons for a "no" will need to be provided. This is the "dragon's den." Managers who are affected by the decision and are not part of the senior team must be a part of the town hall meeting. They must under no circumstances feel overruled or bypassed.

Execution

You can then focus on your primary choice and engage in some rapid innovation to gain an early win. An action plan is drafted during the "I" day.

Implementation of the decisions is ensured by a series of 30-, 60-, and 90-day reviews to confirm that you implement the decisions you have made.

The sponsor ensures that the project teams are adequately resourced and that no subtle sabotage is taking place. The sponsor also has a key task in ensuring that recognition of successful progress takes place. Another key role for the sponsor is facilitating communication.

After the town hall meeting the sponsor starts every management meeting with a report on innovation progress. The first update is two weeks into the project.

This is not an alternative to culture change and a fully developed innovation process. It is the initiator of both of those activities. It will help you find believers and start you on the road to building critical mass.

BROWSER'S BRIEFING CHAPTER 19

- Using change agents is a proven way of initiating the innovation culture change.
- A representative group of change agents needs to be carefully chosen.
- Create a road map and a project plan for change.
- Focus on a product, service, or market segment with declining revenue to gain attention for the need to change.
- An early win is essential.
- Create a "one minute message" to communicate the need for change.
- The business leader is responsible to communicate the vision of innovation.
- The business leader must ensure that people are given time and that roadblocks are removed.
- Innovation must be integrated into the business strategy.
- The business leader must be high-profile in recognizing innovative behavior.
- The "I" day is a high-focus method of initiating the journey to becoming an innovative organization.

20
Creating the Future

The best way to predict the future is to create it.

—Peter Drucker

Innovation has moved up everyone's agenda dramatically in the last five years. This is because we are all becoming increasingly concerned about the future. This is why we are the "innovation generation." We all see the globalization that has been gathering speed over the last 50 years, which shows itself in high-speed communications, an increasing and aging population and, sadly, the deterioration of the natural habitat we call planet Earth.

Happily, there has so far been no limit to human ingenuity in addressing these issues.

Innovation needs people with passion. I invite you to "step out of the box" and release your natural passion. The previous chapter was very much about process, this last chapter provides that ever important balance with people.

Peter Drucker's saying is so true, "The best way to predict the future is to create it." This book has been about how you and I will create that future.

Innovation is about answering the needs of tomorrow. It is not about technology, it is about the needs of people like you and me, and technology is only one way of answering those needs. It is need, not technology, that drives innovation.

There are other myths in the world that belie the truth of innovation. It is the "magic touch" between you and others that leads to radical innovation. We understand far better today the importance of sharing collective knowledge. We understand far better the importance of diversity and the importance of giving ourselves time and space to share knowledge.

The current myth is that the greatest challenge for innovation is that of creativity. In truth, the greatest challenge is getting the new idea "out there." That is because when it is an idea it is not yet a product or service. The organizations we have created all too often get in the way of execution. Lack of organizational agility is the greatest barrier to innovation.

SO WHAT DO WE DO?

The last two decades of the 20th century saw people understanding far better how groups of people, we call them organizations, work well together. The primary symptom of a dysfunctional organization is the failure to meet the customer's needs and expectations—the failure to deliver quality. This led to the emergence of quality management as a way to make organizations more effective at delivering quality.

Our problem is that many quality managers have become focused on correcting the problems of yesterday's failed delivery. They have taken their eye off the ball, and that ball is the customer needs of tomorrow.

This is like a racing driver who keeps tinkering with the engine of the car but never actually takes part in the race.

In the face of global trends it is tempting to think there is little we can do to address the consequences of our mushrooming and aging population, global warming and pollution, and the equalization of the world's wealth.

The temptation is to be like an ostrich, ignore the realities of the world, and keep tinkering with the quality management system. It's time to stop tinkering and take the vehicle out on the road. You have seen that an effective QMS is fundamental to successful innovation, so let's move on and join the race and innovate.

FULFILLMENT FOR THE INNOVATOR

Far too much of business has become obsessed with the score instead of the game. If you play golf or any other sport you know that obsessing about the score will destroy your game whereas focusing on the game will bring about the result you want.

The result we want is a strong and healthy organization with a long-term future, and we know that the game is fulfilling the needs of both today's and, importantly, tomorrow's customer.

Playing a good game is one of the most personally satisfying things anyone can do, and that game can be any activity that we engage in. Innovation, at whatever stage you participate, is one of the most satisfying and fulfilling of all "games."

WHERE DO I BEGIN?

You start with a personal assessment, and then follow it with an assessment at an organizational level.

For successful innovation you must know the aptitudes of the people in your organization. You already know whether you are a creator or connector, a developer or a doer. But what of your colleagues, where are their aptitudes? At a personal level you should also start to develop and operate your network. Find out the aptitudes of other people in your network by asking them to complete the self-assessment and explain to them what it means.

Engage your people where their best contributions can be made and develop the "new behaviors" I have described. These new behaviors release knowledge. You need to release the *wisdom of crowds* through *communities of innovation* in which a key feature is dialogue. This engages the tacit knowledge in your colleagues, which in turn develops your ability to innovate. You can then develop more communities of innovation to build your critical mass as you move toward an innovative culture.

Looking at your organization, an in-depth assessment going deeper than the assessment in this book will point you to where your issues lie and the actions you need to take. You need to look at your own organization and decide where it needs to change and how to make those changes happen. You now have an understanding of the innovation process and the different roles in the process. You should address your innovation process in detail using the assessment tool in Chapters 6 through 10. Do this in discussion with your colleagues, not in isolation. This will point you to some immediate actions you can take with your innovation process. You may well find that you need to look afresh at how marketing, research, development, operations, and sales are structured in your organization.

Moving to a higher level, the innovation process of the organization needs to be integrated with the strategy of the organization. Strategy is about finding "white space" and about engaging in open market innovation. The third aspect of strategy must be the development of the innovative culture and those critical skills and behaviors in the early stages of the innovation process.

Culture and behavior have been addressed in major part in this book. At a strategic level the critical issue is endorsing the behaviors you wish to encourage in recruitment, personnel assessments, and through recognition.

Looking to the future, one issue that is often overlooked with culture change is the "rite of passage." I have not dwelt on this but it should not be overlooked. Letting go of the old and embracing the new is something that cultures through the ages have recognized as essential, whether it is the coronation or inauguration of a new head of state or the coming of age of a member of our own family. This is not just a celebration of the new, this is also a release of the old ways, and this must happen as you move to an innovative culture. Time scale? Think 18 months down the road for your "innovation celebration."

THE RESULT

Self-fulfillment, whether at a personal or an organizational level, is the best thing that can happen in life and yet it eludes the majority of people and the majority of organizations. Ask Maslow! Innovation enables that fulfillment better than any activity you will engage in.

We all have a creative need and yet the needs of the modern organization have all too frequently subverted this personal need. Business is not a battle; it should be about collaboration, not conflict, and be about growth, not greed.

Follow the path I have described and you will become both an individual and an organization in which the left and the right brain are balanced, where there is trust and honesty and you will achieve fulfillment.

WHAT DO I DO NOW?

Know yourself and know your colleagues. Know your organization, use the self-assessment. Do the process and organizational assessment.

Educate and explore. Do a one-day workshop so that you can interact with others who are interested in innovating. Select those new places to go and new people to meet. Take notes.

Get an early win. For yourself and your organization. I already talked about how to do this for your organization. But what about yourself? Realizing your personal potential is essential. Use your own creativity in order to do that. Your early win will involve you doing something you have never done before and maybe were frightened to try.

Explore, observe, take notes, share ideas, and then try that something new. Don't be afraid to fail. Whether you succeed or fail, congratulate yourself for doing something you have not done before. Give a gift to yourself. The gift to yourself should be something that you will remember whether it is an experience or something you can use. The fulfillment when you try something new and succeed will be an excitement that spurs you to try other new things.

I invite you to step out of the box and realize your own potential as an innovator. Join the "innovation generation."

Endnotes

Chapter 1

1. Michael Porter, *Creating and Sustaining Performance by Competitive Advantage* (New York: The Free Press, 1980).

Chapter 2

1. Frank Zollner, *Leonardo da Vinci* (Cologne: Taschen, 2006).
2. Herve Mignot, leader of the French delegation to ISO/TC 176, in conversation with the author.
3. Albert-László Barabási, *Linked* (London: Penguin, 2002).
3. Peter Merrill, *Do It Right the Second Time* (Portland, OR: Productivity Press, 1997).
4. Ikujiro Nonaka and Hirotaka Takeuchi, *The Knowledge Creating Company* (New York: Oxford University Press, 1995).
5. Ross Dawson, *Developing Knowledge-Based Client Relationships* (Oxford: Butterworth, 2000).
6. James Surowiecki, *The Wisdom of Crowds* (New York: Doubleday, 2004).

Chapter 4

1. Peter Senge, *The Fifth Discipline* (New York: Doubleday, 1990).
2. International Organization for Standardization, *ISO 9000:2005, Quality management systems—Fundamentals and vocabulary* (Geneva: ISO, 2005).
3. IBM, *Expanding the Innovation Horizon: 2006 Global CEO Study* (Somers, NY: IBM, 2006).

Chapter 5

1. James Nesbitt, *Megatrends* (New York: Warner, 1982).
2. Club of Rome, *Limits to Growth* (Washington, D.C.: Potomac Associates, 1972).

3. W. Chan Kim and Renee Mauborgne, *Blue Ocean Strategy* (Boston: HBR Press, 2005).
4. Michael Porter, *Creating and Sustaining Performance by Competitive Advantage* (New York: The Free Press, 1980).
5. Ibid.
6. Jim Collins, *Good to Great* (New York: HarperCollins, 2001).

Chapter 6

1. Keith Sawyer, *Group Genius* (New York: Basic Books, 2007).

Chapter 7

1. Jacob Goldenberg and Roni Horowitz, "Finding Your Innovation Sweet Spot," *Harvard Business Review* (March 2003).
2. Stephen Karcher, *The I Ching: The Book of Changes* (London: HarperCollins, 1997).
3. Dennis Snow and Teri Yanovitch, *Unleashing Excellence* (Sanford, FL: DC Press, 2003).

Chapter 8

1. Malcolm Gladwell, *The Tipping Point* (New York: Little, Brown & Co., 2000).
2. Aaron De Smet, Mark Loch, and Bill Schaninger, *Anatomy of a Healthy Corporation* (New York: McKinsey, 2007).
3. George Day, "Is It Real?" *Harvard Business Review* (December 2007).

Chapter 9

1. William Bridges, *Managing Transitions* (Cambridge, MA: Da Capo, 2003).
2. Daniel Kahneman and Amos Tversky, *Choices, Values, and Frames* (Cambridge: Cambridge University Press, 2000).

Chapter 10

1. Anderson, "Customer Value Propositions in Business Market," *Harvard Business Review* (March 2006).

Chapter 11

1. http://en.wikipedia.org/wiki/organizational_culture.
2. Terrence Deal and Allan Kennedy, *Corporate Cultures* (New York: Perseus, 1982).

3. Ibid.
4. Keith Ferrazzi, *Never Eat Alone* (New York: Doubleday, 2005).
5. Peter Merrill, *Do It Right the Second Time* (Portland: Productivity Press, 1997).
6. John Kotter, *Leading Change* (Boston: HBR Press, 1996).

Chapter 12

1. Lynda Gratton and Tamara Erickson, "Eight Ways to Build Collaborative Teams," *Harvard Business Review* (2007).

Chapter 13

1. International Organization for Standardization, *ISO 10015:1999 Quality management systems—Training guidelines* (Geneva: ISO, 1999).
2. Stephen Covey, *Seven Habits of Highly Effective People* (New York: Simon and Schuster, 1989).

Chapter 14

1. Brian Uzzi and Shannon Dunlap, "How to Build your Network," *Harvard Business Review* (December 2005).

Chapter 15

1. Jim Collins, *Good to Great* (New York: HarperCollins, 2001).
2. International Organization for Standardization, *ISO 10006:2003, Quality management in projects* (Geneva: ISO, 2003).

Chapter 16

1. Albert-László Barabási, *Linked* (London: Penguin, 2002).
2. Vijay Govindarajan and Chris Trimble, "Building Breakthrough Businesses within Established Organizations," *Harvard Business Review* (May 2005).

Chapter 18

1. Tuba Ustener and David Godes, "Better Sales Networks," *Harvard Business Review* (July 2006).
2. Larry Huston and Nabil Sakkab, "Connect and Develop—Procter & Gamble Model for Innovation," *Harvard Business Review* (March 2006).
3. James Moore, *The Death of Competition* (New York: HarperCollins, 1996).

Index

A

B

C

D

E

J

K

L

M

N

O

P

Q

R

S

T

U

V

W

X

Y

Z